Living Technology

Living Technology

**Philosophy and Ethics at the Crossroads
Between Life and Technology**

Armin Grunwald

JENNY STANFORD
PUBLISHING

Published by

Jenny Stanford Publishing Pte. Ltd.
Level 34, Centennial Tower
3 Temasek Avenue
Singapore 039190

Email: editorial@jennystanford.com
Web: www.jennystanford.com

British Library Cataloguing-in-Publication Data
A catalogue record for this book is available from the British Library.

Living Technology: Philosophy and Ethics at the Crossroads Between Life and Technology

ISBN 978-981-4877-70-1 (Hardcover)
ISBN 978-1-003-14711-4 (eBook)

Contents

Preface   ix

1. **What You Will Find in This Book?**   **1**
   1.1   The Point of Departure: Observations   1
   1.2   The Mission: Targets and Objectives   4
   1.3   The Way Ahead: Overview of the Chapters   6
   1.4   The Origin: Background   10

2. **Life and Technology as Conceptual Patterns**   **13**
   2.1   Life   14
   2.2   Technology   20
   2.3   Uncharted Territory: Life and Technology
         at a Crossroads   28

3. **Ethics Guiding Our Pathways to the Future**   **35**
   3.1   Problem-Oriented Ethics   36
         3.1.1   Ethics for Analyzing and Resolving
                 Moral Conflicts   36
         3.1.2   Standard Situations in a Moral
                 Respect   39
         3.1.3   Beyond Standard Situations in a
                 Moral Respect   42
   3.2   Ethics for Guiding the Techno-Scientific
         Advance   45
   3.3   Ethics of Responsibility   49
         3.3.1   The Understanding of Responsibility   50
         3.3.2   Consequentialist Ethics of Technology   53
         3.3.3   Responsibility Beyond
                 Consequentialism   56

4. **On the Track to Creating Life: Synthetic Biology**   **61**
   4.1   Synthetic Biology   62
   4.2   The Debate on the Opportunities and Risks
         of Synthetic Biology   68
         4.2.1   Studies on the Opportunities and
                 Risks of Synthetic Biology   69

4.2.2	Overview of Opportunities	72
4.2.3	Categorization of Risks	74
4.3	Ethical Challenges	79
4.3.1	Risks of Self-Organizing Products of Synthetic Biology	80
4.3.2	The Moral Status of Created Organisms	86
4.3.3	Human Hubris by "Playing God"?	89
4.4	Responsibility Configurations of Synthetic Biology	94
4.5	Beyond Ethics: Visionary Narratives Around the Notion of "Life"	101

5. Animal Enhancement for Human Purposes — **107**

5.1	Technologies for Intervening into Animals	108
5.2	The Semantics of Animal Enhancement	113
5.3	Normative Frameworks and Ethical Challenges	117
5.3.1	Between Human Purposes and Animal Welfare	118
5.3.2	Ethical Spotlights on Animal Enhancement	122
5.4	Responsibility Considerations	129
5.5	Beyond Ethics: Changing Human–Animal Relationships	133

6. Shaping the Code of Human Life: Genome Editing — **137**

6.1	Editing Humans? The Case of the Chinese Twins	138
6.2	Genome Editing: Turbo Genetic Engineering	142
6.3	Ethical Issues of Human Genome Editing	147
6.3.1	The Consequentialist View of Germline Intervention	148
6.3.2	Deontological Arguments	155
6.4	Responsibility Challenges of Germline Intervention	158
6.5	Genome Editing: Editing Humans?	166

7. Human Enhancement — **169**

| 7.1 | Technologies for Human Enhancement | 169 |
| 7.1.1 | Converging Technologies for Improving Human Performance | 170 |

7.1.2 Visions of Human Enhancement 174
7.1.3 Enhancement in Human History 180
7.2 The Semantics of Human Enhancement 183
7.2.1 Enhancement *versus* Healing 183
7.2.2 The Semantic Field of Healing,
Weak and Strong Enhancement 187
7.2.3 Human Enhancement by Technology 192
7.3 Ethics of Human Enhancement 195
7.3.1 Normative Uncertainties and Ethical
Questions 195
7.3.2 Ethical Argumentation Patterns 199
7.3.3 Balance of the Ethical Debate So Far 206
7.4 Are We Heading Toward an "Enhancement
Society"? 209
7.5 Responsible Human Enhancement 216
7.5.1 Responsibility Constellation 217
7.5.2 The Case of Neuro-Electric
Enhancement 221
7.6 Changing Relations Between Humans and
Technology 225

8. Robots and Artificial Intelligence: Living Technology? 229
8.1 Artificial Companions 230
8.2 Autonomous Technologies 235
8.2.1 Autonomous Technology in Practice
Fields 237
8.2.2 Autonomous Problem-Solving and
Machine Learning 242
8.3 Ethical Issues 246
8.3.1 Ethical Dilemma Situations 247
8.3.2 Adaptation and Control 250
8.3.3 The Moral Status of Autonomous
Technology 255
8.4 Responsibility Considerations 259
8.4.1 Responsibility Assignment to
Autonomous Technology? 259
8.4.2 Human Responsibility for
Autonomous Technology 267
8.5 Autonomous Technology: Loss of Human
Autonomy? 270

9. On the Future of Life   **277**

    9.1   The Ultimate Triumph of the Baconian
            Homo Faber?   278

    9.2   The End of Nature (1): Human Impact on
            the Future of Life   281

    9.3   The End of Nature (2): Dominance of the
            Techno-Perspective   284

    9.4   Does Autonomous Technology Live?   288

    9.5   Moral Status for Engineered and Created "Life"   293

    9.6   The Ladder of "Life"   298

Bibliography   305

Index   337

Preface

The boundaries between inanimate technology and the realm of the living become increasingly blurred, e.g., in synthetic biology, genome editing, human enhancement, germline intervention, artificial intelligence, and robotics. Deeper and deeper technological interventions into living organisms are possible, covering the entire spectrum of life from viruses and bacteria to mammals, including humans. Simultaneously, digitalization and artificial intelligence (AI) enable increasingly autonomous technologies. Inanimate technologies such as robots begin to show characteristics of life.

Contested issues pop up, such as the dignity of life, the enhancement of animals for human purposes, the creation of designer babies, and the granting of robot rights, calling for philosophical and ethical scrutiny and orientation. The book addresses the *understanding* of the ongoing dissolution of the boundaries between life and technology, the provision of *ethical guidance* for navigating research and innovation responsibly, and the philosophical *reflection* on the meaning of the current shifts. My hope is that it will enrich and strengthen ongoing debates on the future of life, of technology, and of us humans.

I would like to express warmest thanks to my colleagues at the Institute for Technology Assessment and Systems Analysis (ITAS) in Karlsruhe and at the Office of Technology Assessment at the German *Bundestag* (TAB) for many discussions. I am also indebted to Julie Cook and Sylke Wintzer for their great assistance in the linguistic improvement of the manuscript and for taking care of the many references and checking their consistency.

A. Grunwald

Karlsruhe, 6 January 2021

Chapter 1

What You Will Find in This Book?

This brief introductory chapter offers a guided tour through the book. Starting with the basic observations (Sec. 1.1) and unfolding its objectives (Sec. 1.2), it provides an overview of the nine chapters (Sec. 1.3) and their background (Sec. 1.4).

1.1   The Point of Departure: Observations

The scientific and technological advance has made the traditional boundary between inanimate technology and the realm of the living more permeable. Developments in, e.g., synthetic biology, genome editing, human enhancement, and germline intervention demonstrate huge advances in science and technology. Deeper and deeper technological interventions into living organisms are possible, covering the entire range of life, from bacteria through plants and animals to humans. Living organisms are increasingly regarded as the point of departure to investigate and understand them at the micro-level, to modulate them as technically functioning systems, and then to rebuild or replace natural processes with technical ones, according to human intention. Far-ranging change and manipulation of natural organisms has become possible.

Simultaneously, a complementary development is ongoing. Recent developments in the field of digitalization and artificial

Living Technology: Philosophy and Ethics at the Crossroads Between Life and Technology
Armin Grunwald
Copyright © 2021 Jenny Stanford Publishing Pte. Ltd.
ISBN 978-981-4877-70-1 (Hardcover), 978-1-003-14711-4 (eBook)
www.jennystanford.com

intelligence (AI) make increasingly autonomous technologies possible: robots for customer service, intelligent prostheses, autonomous cars, and social bots on the Internet. In this line of development, traditional technology is empowered by capabilities known only from living organisms: self-organization, recognition of its environment and adaptation to changes, the capability to move in an unknown territory, to develop problem-solving strategies on its own, and to learn from experience. Inanimate technology begins to show properties of life.

Therefore, my *first observation* motivating this book is: We are on the track, metaphorically speaking, to create *living technology and technical life*. The boundaries between life and technology are currently becoming more and more permeable from both sides. We are witnessing deeper and deeper technological intervention in life and are experiencing technology acquiring more and more attributes of life (e.g., Funk *et al.*, 2019; Giese *et al.*, 2014; Lin *et al.*, 2012; Sternberg, 2007).

A plethora of new opportunities opens up, e.g., for overcoming diseases and disabilities but also for applications in energy supply and industrial production, for mobility, and in the military. The scientific and technological advance empowers humans to manipulate living organisms to a formerly unknown extent, e.g., by genome editing or enhancement technologies. Autonomous and intelligent technologies can replace humans for dangerous or boring routine activities, thereby also increasing economic efficiency. New services and systems combining technicalized life and living technology can be designed, serving human needs and creating new markets. The health system, agriculture, production, industry, environmental protection, and other areas will benefit. Expectations are high in many fields.

However, concerns as to the possible risks involved are posed, building on earlier debates on the risks of genetically modified organisms (GMO) and food (GMF) for health, the environment, and related social issues. Some of them touch upon specific issues, while others are visionary and more speculative. Artificially produced or technologically modified life could develop further according to the principles of self-organization and could potentially reproduce and get out of control. Challenges to distributional justice and equity could arise, together with questions about the carriers of responsibility in cases of increasingly autonomous technology. Autonomous robots which are able to move independently in an unknown environment, to learn, and to behave similarly to humans motivate and fuel new debates around responsibility, new human–machine interfaces,[1] control, and the moral status of the robots themselves.

Therefore, the *second observation* guiding this book is the emerging need for orientation in coping with all the new opportunities and dealing with the challenges and risks responsibly. Contested issues such as the dignity of life, limits to enhancing animals for human purposes, interventions into the human germline, control and power, as well as distributional justice with regard to harvesting the benefits of the new opportunities provide a vast field of challenges for responsible deliberation, decision-making, and action. Applied ethics, in particular bioethics, medical ethics, and the ethics of technology are called for (e.g., Comstock, 2000; Buchanan, 2011; Paslack *et al.*, 2012; Gunkel, 2018; Li *et al.*, 2019).

Furthermore, transgressing the boundaries between life and technology is a deep challenge beyond morality and applied ethics. It affects the traditional cultural, religious, philosophical, and ontological order that has emerged over the course of human history. While living beings have traditionally been regarded as having their origin in nature, in evolution, or in God's creation, artificially produced or deeply manipulated living beings would be human creations. In particular, the relations of technology and life, as well as between humans and technology, need to be reflected. Core notions of the human perception of the world, such as "life" and "technology," must be reconsidered.

Accordingly, the *third observation* fueling this book is that dissolving the boundaries between life and technology needs deeper philosophical understanding, beyond applied ethics. Questions about reviewing and sharpening core issues of human perception of the world, and our self-conceptions, lie behind many of the tasks for ethical reflection. Therefore, philosophy of nature, philosophy of technology, and anthropology also have to be engaged in achieving a better understanding of the ongoing fundamental changes in human civilization driven by science and technology (e.g., Winner, 1982; Glees, 2005; Habermas, 2005; Sandel, 2007; Savulescu and Bostrom, 2009; Hurlbut and Tirosh-Samuelson, 2016).

---

[1]Throughout this book, gender neutral and non-binary language will be used.

1.2   The Mission: Targets and Objectives

The diagnosis of dissolving or even disappearing boundaries between technology and life as well as the resulting far-ranging orientation needs for science, philosophy, policy-making, and society require clarification and substantiation. According to ethical and philosophical guidance, conclusions must be drawn in order to shape the technological advance and make the best use of it while avoiding unbearable risks. This book follows three major objectives related to the observations mentioned above:

  (1) *Substantiating the diagnosis*: The diagnosis of dissolving boundaries will be underpinned by in-depth case studies. Within these studies, I will analyze various highly dynamic fields, which lie between biology, medicine, and technology, with respect to the process and shape of the dissolution, the state of the art of science and technology, the intended applications and challenges, and the perspectives and visions ahead. This will be conducted in the fields of synthetic biology, animal enhancement, human genome editing, human enhancement, and autonomous technologies. Obviously, a whole book could be dedicated to each of these fields, while the space available in this volume is limited. Therefore, I will focus on those aspects specifically related to the changing boundary between technology and life (e.g., Chopra and Kamma, 2006; Tucker and Zilinskas, 2006; Boogerd *et al.*, 2007; Adamatzky and Kosiminski, 2009; Gutmann *et al.*, 2015).

  (2) *Providing ethical guidance*: In order to meet the demand for orientation, ethical guidance for navigating responsibly into and through the age of *living technology* will be provided. While the scientific and technological advance offers a vast range of new opportunities, it simultaneously leads to debates, controversies, and even hard conflicts with two different origins. First, unintended side effects have accompanied the entire techno-scientific advance (Grunwald, 2019a), which applies to the fields considered in this book, too. Uncertainty and risk are therefore an inherent part of their development as well as of public and scientific concern. Second, the moral status of living organisms, e.g., animals, could be affected

or even violated by technical intervention, e.g., by animal enhancement or human genome editing. The philosophical approach to responsibility will be applied as the overarching concept to make ethical reflection in both directions operable and to support the creation of responsible pathways to the future (e.g., Miller and Selgelid, 2006; Merkel *et al.*, 2007; Owen *et al.*, 2013; Cussins and Lowthorp, 2018; Nuffield Council on Bioethics, 2018).

(3) *Reflecting anthropological shift*: While ethical and responsibility reflections are without any doubt needed to provide orientation for action and decision-making for the next steps toward the future, they cannot cover all the demands for orientation. New interfaces and crossroads between life and technology and the transgression of the former strict boundary affects human perception and conceptualization both of the outer world and of human self-images. This book, therefore, is also dedicated to reflecting the concepts of life and technology, including their generally strict separation, at least in Western reasoning. Because these have deeply shaped human mind and thought for centuries and even millennia, new accentuations or shifts of meaning are of the highest importance and have to be investigated. To this end, philosophy of nature, philosophy of technology, and anthropology will be involved, beyond ethics (e.g., Searle, 1980; Grunwald *et al.*, 2002; Reggia, 2013; Wallach and Allen, 2009).

While many books and edited volumes are available dedicated to specific areas of the dissolving boundaries and newly emerging interfaces between life and technology, this book offers three specific perspectives, conceived along the observations it is based on:

- Most of the books available look either at technical interventions into living systems, such as animals or humans, or at technical artefacts like robots receiving more and more autonomy, which was formerly a property of life only. Considering and reflecting the dissolution of the boundary between technology and life *from both sides*, as is done in this book, is specific.

- While ethical inquiry in the fields touched upon sticks to the respective sub-disciplines of applied ethics, such as bioethics, animal ethics, ethics of technology, or risk ethics, this book goes further. It provides deeper scrutiny with respect to philosophy and anthropology in order to achieve better understanding of the dissolution processes at the boundaries between life and technology and the emergence of new crossroads.
- Usually, books focus on specific areas of life, e.g., on gene editing of plants, on interventions into the human germline, or on artificial intelligence showing some properties similar to life. This book, however, presents a comprehensive look at the entire spectrum of living organisms: bacteria and viruses, plants, animals and humans – and robots, as possible early forms of emerging technical life.

This approach aims to provide new philosophical and ethical insights in view of the dissolution of the boundaries between life and technology as well as newly emerging opportunities and risks. Careful reflection, deliberation, and action will be of major significance for shaping the future of humankind and the ecosystems in the Anthropocene.

1.3 The Way Ahead: Overview of the Chapters

The book follows a simple structure. Chapters 2 and 3 set the stage by preparing the notions, tools, and argumentative patterns needed for the subsequent ethical and philosophical inquiry. Chapters 4 to 8 comprise the case studies on synthetic biology, animal enhancement, human genome editing, human enhancement, and autonomous technology. Each of them includes a subchapter dedicated to specific responsibility issues. Chapter 9 reflects the case studies with respect to overarching philosophical questions at the blurring interfaces between life and technology and emerging new configurations.

Chap. 2: Life and Technology as Conceptual Patterns

The diagnosis motivating this book, of dissolving boundaries between life and technology, needs clarification. For preparing the case studies, which aim to provide deeper insights, it is necessary to introduce both major notions, *life* and *technology*, with respect to their meaning. Because this meaning is, in both cases, controversial and subject to different perspectives, a brief but concise explanation is needed in order to be able to use them in the case studies in a transparent manner. Beyond the introduction to these notions for the entire book, a first and still merely conceptual view is provided of the currently dissolving boundary between them and of emerging demands for orientation.

Chap. 3: Ethics and Philosophy as Guiding Reflection

The scientific and technological advance raises various questions with regard to responsibility. While this applies to almost all fields of the advance, such as energy technologies, digital services, and nanotechnology, specific challenges emerge as soon as life is touched. In this chapter, the approach to ethics and the philosophy of responsibility will be presented for application to the subsequent case studies. Applied ethics has developed into a huge field of ethical reasoning over recent decades. However, reflecting new and envisioned technical and scientific opportunities for transgressing the formerly strict boundary between technology and life in this volume needs to involve not only the ethics of technology but also bioethics and environmental ethics. While responsibility is usually rooted in consequentialist ethics, it will also be necessary to go beyond consequentialism and involve other sub-disciplines of philosophy, such as anthropology, philosophy of life, and the philosophy of nature.

Chap. 4: On the Track to Creating Life: Synthetic Biology

Huge advances in nanobiotechnology and systems biology, increasingly benefitting from the potential of ongoing digitalization, have fueled the emergence of synthetic biology. High expectations of humans becoming able to create living systems from scratch in some future have been raised, at least at the level of microorganisms. This field has motivated heavy debates on responsibility and its distribution over different actors. Simultaneously, it quickly became clear that synthetic biology is not just another sub-field of biotechnology but rather has the potential to change our view on life as such. At its early stage, notions such as "engineering life" and "evolution 2.0," determined by humans, already shed light on a possible paradigmatic change in human perception of life. The task of this chapter is to review the state of the art of the responsibility debates on synthetic biology and to extend them by applying the framework of problem-oriented ethics and related responsibility. It will include a systematic elaboration of the ethical issues related to possible risks as well as concerning the moral status of created organisms. Beyond applied ethics, the changing relationship between technology, nature, and life will be considered, in particular with respect to challenges for the notion and understanding of life itself.

Chap. 5: Enhancing Animals for Human Purposes

During human history, animals have been used for human purposes, for food supply and farming, as companions, for hard work, for sports and the military, and, more recently, for scientific experiments. Keeping and breeding animals were early forms of enhancing them in the sense of making animals better fit human purposes. Genetic modification opened up the pathway for deeper and faster intervention into animals since the 1970s. More recently, animal enhancement builds on converging technologies, nanobiotechnology, and gene editing to further improve the usefulness of animals for human purposes. This chapter describes some of the key directions of research that are currently in progress in this field, identifies emerging ethical challenges, and reflects on the responsibility distribution of animal enhancement. Final considerations lead beyond the level of applied ethics to the hermeneutics of the changing relationship between humans, animals, and technology.

Chap. 6: Shaping the Code of Human Life: Genome Editing

Genetics and genetic engineering have been developed since the 1960s, following the discovery of the double helix DNA structure in 1953. The development of these technologies was and still is accompanied by partially heavy debates on ethical issues. Parts of these debates are motivated by concerns about possible risks to human health and to the natural environment, while other parts address more fundamental issues of the moral status of affected life. The most intense ethical debates address gene technologies applied to humans, usually dedicated to medical purposes and to the area of reproduction. The most recent wave of ethical reasoning in this field followed the availability of the CRISPR-Cas9 genome editing technology, which makes interventions into genomes cheaper and more effective. In this chapter, the focus is on genome editing applied to humans, in particular on germline interventions. Taking the story of the Chinese researcher He Jiankui as an illustrative example, I will analyze normative uncertainties and ethical issues and the resulting issues of responsibility. In this case, the ethically most challenging issue is the possible impact on future humans by intervening into the human germline of those who cannot give their informed consent. Finally, I will raise the question whether genome editing corresponds to editing *humans*.

Chap. 7: Human Enhancement

A controversial international debate about enhancing human beings by new technology has been ongoing since the beginning of the 21st century. Building on previous considerations of improving humans by genetic engineering or eugenics, this more recent wave occurred in the course of the so-called converging technologies. The debate on human enhancement enriched and fueled intellectual reasoning about the future of humans by adding further ethical, anthropological, and philosophical aspects, in particular in the fields of post- and transhumanism. This chapter introduces the main ideas of the underlying enhancement stories, including the visions and utopias presented. A crucial issue is to clarify the semantics of technologically "enhancing" humans, because this provides several insights for performing the ethical investigation. The major concerns raised in the ethical debate so far will be presented and critically discussed, resulting in a diagnosis of the absence of strong ethical arguments against human enhancement *per se*. Conclusions for the responsibility distribution will be drawn. Finally, the changing relations between humans and technology in the course of human enhancement are investigated.

Chap. 8: Robots and Artificial Intelligence: Living Technology?

Recent developments in the field of digitalization and artificial intelligence (AI) make increasingly autonomous technology possible. Many developments and implementations of autonomous technology are ongoing in fields such as industrial production, autonomous driving, and care robots. As soon as this technology, e.g., an autonomous robot, is able to behave independently in an unknown environment, to learn, and to behave similarly to living organisms, in particular humans in some respect, new debates around responsibility, human–machine interface, control, and the moral status of the robots emerge, as can already be witnessed. The question whether and in what respect these technologies could be considered as some form of life in future is still open. The narrative of artificial companions, considered analogously to human companions, serves as an illustrative door opener, telling us a lot about current expectations. The resulting major normative uncertainties and ethical issues will be considered, which give rise to reflecting responsibility in this highly dynamic development. Finally, the question whether increasing technological autonomy made feasible by digitalization and AI could decrease human autonomy will be tackled.

Chap. 9: On Futures of Life and Technology

The final chapter will harvest the insights gained in the case studies and reflect them at the more general level of relations between life and technology. Major results are the recognition of powerful narratives dominating current perceptions and concerns, the turn from an ontological consideration of the relation between life and technology to a more epistemological and methodological perspective and the finding that the use and understanding of models, primarily the model of a machine, is at the core of different positions and many misunderstandings. To seek better understanding of the changes at the interface between life and technology, human patterns of recognition, as well as anthropomorphic and technomorphic language have to be reflected in a better and more transparent manner.

1.4 The Origin: Background

This book is, in parts, a second and widely modified and modernized edition of the book "Responsible Nanobiotechnology: Philosophy and Ethics" (Grunwald, 2012). While it continues to apply the framework of problem-oriented ethics and responsibility (Chap. 3), the character of the book has changed considerably from focusing on the highly specific field of nanobiotechnology to considering paradigmatic changes at the interface of life and technology in various fields of research and innovation.

The chapters on synthetic biology (Chap. 4), animal enhancement (Chap. 5), and human enhancement (Chap. 7) correspond to chapters in the 2012 book. However, because the focus has changed, they not only had to be adapted to the scientific and technological state of the art in 2020 but also had to be rewritten and complemented with respect to the new focus. In order to complement these considerations, two new case studies on human genome editing (Chap. 6) and on autonomous technology (Chap. 8) were added. The new Chapters 2 and 9 provide the framing of the book: while the basic concepts of *life* and *technology* will be introduced and clarified in Chapter 2, the overall reflection on all of the case studies will be provided in Chapter 9, to reach general conclusions.

All the case studies benefitted substantially from projects conducted at the Institute for Technology Assessment and Systems Analysis (ITAS) at Karlsruhe as well as at the Office of Technology Assessment at the German *Bundestag* (TAB). I would like to express warmest thanks to my colleagues for many discussions on the issues of synthetic biology, animal enhancement, human genome editing, human enhancement, and autonomous technology.

Chapter 2

Life and Technology as Conceptual Patterns

The diagnosis of dissolving boundaries between life and technology, which motivates this book, needs clarification. First, we have to ask about the very nature of the boundary between technology and life – only if we have an idea of this boundary, will we be able to understand what its "dissolution" could mean. To this end, it is necessary to introduce the two major notions: *life* (Sec. 2.1) and *technology* (Sec. 2.2). In both cases, their meaning is controversial and subject to various perspectives and interpretations; thus a concise explanation is needed in order to use them transparently in the subsequent case studies.

This must be done appropriately, by observing the communication situations in which these notions are used. By investigating and debating relations between life and technology, we presuppose that we can talk about them on an equal footing in linguistic and semantic respects, e.g., at the same level of abstraction. Therefore, we have to determine the level of abstraction, which, on the one hand, represents the linguistic phrase "life and technology" and, on the other, allows substantial insight into relevant aspects and facets of both terms. To do this, I will introduce and clarify both notions as *concepts of reflection*, according to philosopher Immanuel Kant (following Hubig, 2007; Nerurkar, 2012; Nick *et al.*, 2019).

Living Technology: Philosophy and Ethics at the Crossroads Between Life and Technology
Armin Grunwald

Copyright © 2021 Jenny Stanford Publishing Pte. Ltd.
ISBN 978-981-4877-70-1 (Hardcover), 978-1-003-14711-4 (eBook)
www.jennystanford.com

Usually, we apply terms to order things or perceptions in a logic of subsumption, subordinating individual things under a general predicate. The predicate "furniture," e.g., comprises things like chairs, tables, and wardrobes; similarly, the predicate "primate" assembles chimpanzees, orangutans, and gorillas, according to rules of subsumption valid in the systematics of zoology. According to predicate logics, a primate is a gorilla, or it is not. Therefore, the respective predicates allow us to bring order to the world surrounding us and make communication and cooperation easier. The introductory claim in this chapter is that the terms *technology* and *life* are also predicates but with a different character.

Concepts of reflection do not contribute to an *ontology* of the world by classifying things and subsuming them under strict and well-defined categories, but rather are, according to Kant, patterns of our recognition of the world, which need reflection. "Life" as well as "technology" are not objects belonging to the empirically researchable world. They rather describe human patterns to reflect on issues in the world from different perspectives and with regard to different cognitive interests. This may sound complicated, and perhaps too sophisticated to some. However, we will see throughout this book the significance of this determination. First, however, this approach will be explained in-depth for both *life* (Sec. 2.1) and *technology* (Sec. 2.2), demonstrating that *life* as a concept of recognition does not live and *technology* cannot be used as an instrument.

Based on these considerations, which go back to various traditions in philosophy, we can give a preliminary response to the question of the nature of the boundary between technology and life (Sec. 2.3), which allows a more precise preparation of the questions for the subsequent case studies.

2.1 Life

The term "life" belongs to the core concepts of anthropology, biology, and philosophy as well as to everyday and common language (e.g., Budisa, 2012; Brenner, 2007; Gayon, 2010; Cleland, 2019). It is, like many core concepts, semantically controversial. Life as the form in which living beings exist (Gutmann, 2017) and develop in nature, different from technology, is only partially accessible to human influence but also shows its own dynamics. Understanding life and assigning meaning to this notion is open to numerous interpretations in culture, religion, philosophy, and science.

When talking about science, the life sciences in general and biology in particular immediately come to mind, because biology as *logos* of *bios* traditionally claims to be *the* science of life. However, the question of the meaning of "life" does not find a clear answer in biology. The encyclopedia *Wikipedia* notes:

> There is currently no consensus regarding the definition of life. One popular definition is that *organisms* are *open systems* that maintain *homeostasis*, are composed of *cells*, have a *life cycle*, undergo *metabolism*, can *grow*, *adapt* to their environment, respond to *stimuli*, *reproduce* and *evolve*. Other definitions sometimes include non-cellular life forms such as *viruses* and *viroids* ("Life." *Wikipedia*. https://en.wikipedia. org/wiki/Life; accessed 13 Apr. 2020).

The absence of a consensual understanding of what is regarded as life is frequently not even perceived as a problem: "[...] the impossibility of a sharp distinction between animate and inanimate would not create difficulties for biology in its everyday scientific practice" (Budisa, 2012, 101; cp. also Machery, 2012). In spite of biology claiming to be the science of life, it does not own a clear and uncontested understanding but rather operates with a plurality of approaches. This seemingly does not create problems – perhaps this variety is even fruitful in many cases (Funk *et al.*, 2019). By going deeper into the philosophy of biology, it becomes clear that it is not "life" *as such* that is the object of biology but rather *living organisms* (Moreno, 2007; Nick *et al.*, 2019; Gutmann, 2017):

> [...] life will be no longer a concept for the natural sciences, but just a convenient word in practice, in the world we inhabit. "Life" will be a folk concept. Its specialists will be no longer chemists, biologists, and roboticists; life will be a subject for psychology, cognitive science, and anthropology (Gayon, 2010, 243).

But what can we learn about the general notion of life? If we want to discuss issues at the boundary between life and technology, we should have an idea of the meaning of "life" and "technology" in a general sense. Knowledge about living organisms or technical artefacts is fine but does not suffice.

As a folk concept (cp. quote above), however, life is at risk of serving as a mere buzzword: "[...] the claim of 'making life' [in synthetic biology, cp. Chap. 4] is more for the sake of gaining attention and grants than for the purpose of formulating a scientific goal" (Funk *et al.*, 2019, 181). Therefore, we cannot rely on a folk concept of life for undertaking our analysis and reflection. In the following, I will explain the understanding of "life" applied in this book, mainly based on Gutmann/Knifka (2015) and Nick *et al.* (2019), starting with a linguistic consideration.

The word "life" is used in colloquial language as a more or less self-evident or self-explanatory notion, in spite of the various and partially contested perceptions in biology and philosophy (Cleland, 2019). In colloquial language, "life" is a noun claiming to describe an existing object: *the* life. Attempts to explain nouns usually seek answers to "what is" questions such as "What is life?": "[...] the grammatical status of the noun produces the *pure impression* of reference as in 'this table' or 'this glass'" (Gutmann and Knifka, 2015, 58). "What is" questions for the meaning of life are based on an *ontological* presupposition: that life is a part of the objective world, clearly separated from other, non-living parts. The question reads as: "What is it *really* that we subsume under the label of life (analogously the term 'technology,' cp. next section)?" This assumption implies that this separation objectively belongs to the world as it *really* is, instead of being a result of human perception, distinction, and recognition.

Many philosophers, following Immanuel Kant (e.g., Hubig, 2007; Nerurkar, 2012), emphasize that attempts to respond ontologically to "what is" questions at this general level will necessarily lead into severe logical and semantic problems. For example, life obviously must have relations to living entities but does not live. The reason is, simply speaking, that life as a general notion cannot be explained by referring to its defining attributes (*definiens*) without a further step, as is possible for furniture and glasses, for example. This further step consists of *reflection* in addition to simple subsumption. Instead of trying to characterize *the* life as an object like a glass or table (cp. quote in the previous para), we will take a different approach and consider "life" as a term of reflection: "As a consequence, 'life' shares its methodological status with other expressions of this kind such as 'atom' or 'time'" (Gutmann and Knifka, 2015, 58). This shift of consideration needs explanation.

In the framework of the linguistic turn proposed by Ludwig Wittgenstein, "life" first is a word with four letters (in English) used in daily and scientific language. We can explore its meaning by investigating how the term "life" is used in communication, how well-functioning communication and cooperation make use of it, and which failures and misunderstandings occur, and for what reasons. Doing this, however, does not suffice to avoid the problem of an ontological misunderstanding, which arises from the "what is" type of question. In order to circumvent this question typical to nouns, it is preferable to start reflecting on the meaning of "life" by looking at how we use the verb "to live" or adjectives such as "lively" or "living" (Nick *et al.*, 2019).

We talk about living entities in language by calling them lively and, thus, making a distinction between lively and non-lively. Regarding an orange tree or looking at a fox lying on the street we have evident and consensual indicators of what we regard as lively: If the tree greens and blooms in spring and produces fruit, we will assign the attribute "lively." If the fox on the street no longer breathes then we will assume that it is dead, i.e., no longer living. At the level of specific objects, we do have an understanding of what "living" means. Moving from the level of the adjective "lively" or "living" to the active verb "to live," we can see that the verb "to live" is a second-order concept (Gutmann and Knifka, 2015): we say that an object lives if we have evidence to assign the attribute "lively" to it (Nick *et al.*, 2019). Generalizing this step leads to the explanation of "life": life as a concept of reflection assembling all the ideas and perceptions we have about living objects and which we regard as specific when talking about them as living objects. The essential mode of operation of this type of concept is *that we apply a specific perspective* while forming an abstract concept such as "life" from daily observation (Gutmann and Knifka, 2015).

The singular *"the* life" is therefore a sensitive and sometimes even problematic notion, which can easily be misunderstood in the sense of an ontological object being part of our world, assuming that all the objects in the world separate into living or non-living ones. Even if this cannot be excluded – it may be the case – we should be modest and not ignore the fact that our patterns of recognition play an important role here. Indeed, we create the meaning of "life" by applying a specific perspective to living objects orientated by a cognitive interest. We could also apply different perspectives, e.g., to cows: we can regard them as possible food, as companion in rural life, or as a kind of machine for pulling the plough in traditional farming. All these and further perspectives do not exclude one another but might depend on the specific communication context and the respective cognitive interest. Take the example of a tree: we can regard it as an element important for the microclimate in a town, as a challenge for climbing, as a source of shade in hot summers, or as a living object. In either case, we apply different perspectives with different opportunities to generalize and to abstract. In the case of regarding a tree as a nice opportunity for learning to climb, we could compare it with rocks or with climbing facilities in a sports hall – which, obviously, is different from tackling it as a living entity together with tigers, bacteria, and humans.

This observation leads to the next question: how can the specific perspective of thematizing entities *as living* be characterized? In this respect, we follow famous biologist Ernst Mayr:

> These properties of living organisms give them a number of capacities not present in inanimate systems: A capacity for evolution, a capacity for self-replication, a capacity for growth and differentiation via a genetic program, a capacity for metabolism (the binding and releasing of energy), a capacity for self-regulation, to keep the complex system in steady state (homeostasis, feed-back), a capacity (through perception and sense organs) for response to stimuli from the environment, a capacity for change at two levels, that of the phenotype and that of the genotype (Mayr, 1997, 22).

This list includes several attributes which, according to Mayr, all living entities share. Following Nick *et al.* (2019, 206), who also refer to Mayr, "life" then can be characterized as the set of activities which objects should be able to perform if being subsumed under *living* or *lively* entities. They should have the capabilities to (without claiming completeness):

- reproduce
- perform metabolism
- grow

- convert and store energy
- perceive environmental conditions and react
- move (internal or external)

By applying the term "life," we ask about the generalizable capabilities living objects should have. Otherwise, they would not be classified as living. Thus, "life" is not an empirically observable part of the world of things but a human pattern of distinguishing living and non-living objects at the general level. This constructivist description avoids artificial problems, e.g., the question of an independent force making living beings live (vitalism), with subsequent ontological or even metaphysical questions. Applying the list mentioned above, or similar lists, allows the introduction of further distinctions and classifications in biology and its sub-disciplines to structure the entire field of living objects.

The scientific view on life needs further clarification. The keyword in this respect is "organism." Biology does not look at life *as such* but at living objects terminologically called organisms (Gutmann and Knifka, 2015; Nick *et al.*, 2019). Exploring the internal functioning of organisms and their external relation to the environment constitutes the epistemological view of biology on living objects (Nick *et al.*, 2019). Thereby, biology operates by presupposing that these organisms are well organized in order to function internally as well as to behave and survive externally. Understanding this organization with respect to the involved processes, materials, flows of material, energy, and information is at the core of biology (Moreno, 2007). This rationale stands behind, e.g., genetics, molecular biology, cell and subcellular biology, while systems biology considers the co-functioning of parts of the system under consideration (Boogerd *et al.*, 2007).

An interesting observation referring to the field of technology is that biology, at least modern molecular biology, follows the model of a machine, implying that biological knowledge has technical roots (Gutmann and Knifka, 2015). Machines consist of many parts, devices, and materials; they convert energy, organize forces, process data, and have moving elements. They can be regarded as an organized system of many parts, which have to co-function in order to serve the overall functions of the machine (Nick *et al.*, 2019). Biology operates in an "as if" mode, making use of the machine paradigm of technology (Gutmann and Knifka, 2015; cp. next section): it analyzes organisms "as if" they were machines without claiming that they *are* machines. In the case of synthetic biology:

> A constructivist – engineering-like – approach is essential for modern scientific thinking and actions. Synthetic biology introduces this epistemological principle into a new realm, the realm of biology: designing life as if it were a man-made machine (Funk *et al.*, 2019, 178).

This perspective allows exploration of causal relations between elements of the considered organism and, thereby, applying the cognitive pattern of causality (following Immanuel Kant) to living objects. This is both a valuable insight per se and an appropriate bridge to the field of technology, as will be shown later.

2.2 Technology

The argumentation in this section follows the same scheme as above, starting by considering the use of the word "technology." In everyday life, communication about technology is usually unproblematic, because a *specific* technology, *specific* techniques, or the attribute "technical" at the occasion of a *specific* setting and framing will be the focus. The frame of meaning is given through the respective context of the communication situation, e.g., in the household, in dealing with the automobile, or in industrial production. In colloquial language and daily communication, we do not have problems or ambiguities in talking about functions or other properties of the technology under consideration, or complaining about malfunction. This also is clear for professional contexts, in particular within the scope of engineering professions, including the activities of teaching, learning, research and development, maintenance and disposal. In any specific context the meaning of technology is clear, and related problems can be solved among the participants by communication and cooperation, e.g., to get malfunctioning technology functioning.

However, as soon as we ask for a more general and even context-invariant understanding of technology by raising the question of the meaning of "technology" *as such* we run into problems. Similarly to the situation of biology with "life," the sciences specifically dedicated to technology – engineering sciences – do not have a consensual and comprehensive definition of technology which covers all areas of engineering and all types of technology. An attempt was made by the German Association of Engineers at the occasion of guidelines on technology assessment (VDI, 1991). Accordingly, technology comprises:

- the set of use-oriented, artificial, concrete objects (artefacts or object systems);
- the set of human institutions in which object systems originate;
- the set of human activities in which object systems are utilised. (VDI, 1991, 2).

This definition fits well with engineering tradition and has several advantages. It looks not only at technical objects and object systems but also includes human actions of making and using technology. It includes characteristic and well-fitting words such as "artificial" and "use-oriented," thereby implicitly referring to Aristotle's distinction between nature as grown and culture as human-made, with technology clearly belonging to the realm of culture. By mentioning "use-oriented," it also refers to the sphere of human intention to make technology, namely to use its functions and to create benefit and utility.

However, this and many other similar definitions have been criticized. In particular, philosophers have made four major points (e.g., Hubig, 2007; Grunwald and Julliard, 2005).

(1) The definition is arranged around technical objects such as machines or their products. The so-called "object systems" (see quote above) are at its core, while also mentioning their production and use. The dimension of technology as *process* is not, or is at least under-represented. For example, it will be difficult to apply the definition to software for computer apps or to logistics of production processes. The restriction to material objects fails to observe the established view of the philosophy of technology subsuming material artefacts as well as procedures and processes under the realm of technology (e.g., Mitcham, 1994).

(2) The definition narrows the realm of technology to *engineered* technology. In everyday life, however, we also speak of technology on completely other occasions. In talking about

piano concerts, the technical quality of the pianist can be an issue. We talk about meditation techniques and mediation techniques, about social engineering, and about the technical precision of a dance group. Even when we do not use the word "technology" here, the words "technique" and "technical" are in close semantic neighborhood.

(3) The definition primarily addresses technology engineered and produced in traditional industry. It excludes, however, the dimension of *technoscience* (Asdal *et al.*, 2007; Nordmann, 2010) as co-developments of classical (natural) science and technology. This notion conceptualizes recent developments in science and engineering as overcoming traditional boundaries between curiosity-driven natural science and application-driven engineering. Most of the case studies analyzed throughout this book can be subsumed to technosciences such as synthetic biology (de Vriend, 2006; Giese *et al.*, 2014; cp. Chap. 4).

(4) The definition presupposes a strict separation between the world of technology – object systems with specific properties – and the rest of the world. Objects not falling under the definition are not regarded as technology. Therefore, the definition has an *ontological* character: it divides the world of things into two species: technology and non-technology. This approach, however, does not meet either cultural and practical experience or linguistic customs. Regard, for example, a piece of nature like the branch of a tree. This is not artificial but can be used *as* technology, for example, for fishing something out of a river. Another example is a small bronze statue. A householder could knock out a thief by hitting them with the statue, thereby regarding it not as a piece of art but as a kind of useful technology.

The fourth point includes the most serious criticism of the above definition. It will be used as point of departure for introducing a different understanding of technology. Behind this point stands the crucial observation that, according to our use of language, technology cannot be a container notion for specific object systems being part of the world while excluding others. Instead, subsuming something under the realm of technology depends on *perspectives and context*.

The notion of technology, therefore, is not an ontological term but necessarily includes a pragmatic dimension related to respective human perspectives and perceptions due to a cognitive interest. At the general level addressed here, "what is" questions should be avoided.

In contrast to the widespread understanding observed also in the VDI definition above, the term "technology" is not a general predicate for the life-world technologies, in the sense that the term "furniture" unanimously subsumes objects such as tables and chairs according to a straight and consensual definition (Grunwald and Julliard, 2005). This ontological perspective misses the *cognitive interest* (Habermas, 1970) people have while talking about technology in general. A thing or object of experience is not just technology or not. Rather, it is regarded as technology in a specific respect, or not. Talking about something *as technology* or discussing *technical* aspects of something constitutes it *as technology* or *as something technical* only, as demonstrated by the examples of the statue and the piano player.

Hence, the concept of technology turns out to be a *concept of reflection* with which humans refer to particular aspects of technology in a generalizing manner. Whoever speaks of technology in general is interested *in certain aspects* of technology with an *intention of generalizability*. By using the general concept of technology, we reflect on different perspectives, among which we determine technical aspects in actions and objects. By talking about technology, implicit or explicit distinctions are applied, e.g., between technologies and elements of the natural environment, or between technologies and pieces of art. According to this understanding, the attribute "technical" becomes an attributive term assigned to an action or an object. Then it is subsumed under the realm of technology according to a certain perspective, analogously to the attribute "living" in the preceding section. Examples of those perspectives are widely-used attributes for technology at the general level, such as (partially also included in the VDI definition above):

- use-oriented
- artificial
- functional
- organized

Four levels of reflection on technology *in general* can be distinguished, without claiming completeness. They all presuppose the Aristotelian view on *techné* as human-made objects for specific purposes (*poiesis*), thus putting technology in any reflective dimension under means/ends-rationality:

(1) Generalizing discourses on technology start, often implicitly, by determining the perspective of the *reflection on general aspects of technologies*, as we did for living entities above. Familiar cognitive interests motivating people to speak of technology in general are distinctions between technology and something different. The distinction between technology and nature is widely used, in prolongation of the Aristotelian tradition. The distinction between technology and art is another which gives rise to many discussions. Usually, such reflections end up in clarifications of the more or less technical aspects of some pieces of art, e.g., an animated movie with innovative tricks, or of technical elements of nature, while, e.g., talking about the hunting techniques of lions. Other perspectives of talking generally about technology include its instrumental character, its relation to humans, and its relation to life, as in the focus of this book.

(2) Another perspective of reflection consists of examining the nature of technology as a *generalized means* (*Inbegriff*). In this case, general functions of technology could be made an issue, e.g., stabilizing society by implementing technical entities with a high reliability and stability such as infrastructures, or contributing to human emancipation from nature due to the European Enlightenment (Francis Bacon). As another possibility, technology in general could be regarded as a compensation for the deficits of humans as imperfect beings (Gehlen, 1986). Transhumanist authors have expanded this perspective to the imperative to making deficient – in their perception – humans superfluous, in favor of an optimized technical civilization.

(3) A third generalizing perspective asks about possible *surplus* of technology beyond the initially determined means/ends-relation (Rohbeck, 1993, with reference to Dewey, 1920). Human creativity in technological thinking allows

for expansion of the possible usage of technology beyond the initial intentions. Inasmuch as an object is described as "technology," the question is not only whether this object is a means to a *specific* end under *clear* conditions but also whether it could be an appropriate means *beyond* and serving other purposes. The idea of technology includes thinking beyond the limitations present in any concrete technological product or process, allowing for creative innovation, but also for dual-use and even misuse.

(4) Finally, and well prepared by the preceding issue, the occurrence of *unintended side effects* of technology could also be made an issue at the general level. According to unforeseen or unforeseeable developments, unintended side effects have been accompanying the technological advance since its beginning. Their occurrence at the more technical level in combination with social effects such as dual- or misuse was among the motivations to characterize modern society as a "risk society" (Beck, 1986), with side effects being an essential part of it (Böschen *et al.*, 2006).

Therefore, the question, "What is technology?" is misleading. It invites ontological answers under a senseless dichotomy: something either is technology or it is not. Instead, the cognitive interest guiding human communication and cooperation is decisive to applying specific perspectives on objects of the world with respect to their technical aspects. The notion of technology is not a container term subsuming all the concrete technologies and techniques but demarcates general perspectives of communication on these technologies related to specific cognitive interests.

In this book, the cognitive interest is about analyzing and reflecting changes at the interfaces and crossroads between technology and life (Chap. 1), so the notion of technology applied must be determined accordingly. This must happen without pre-determining any *results* of the analyses and reflections in the subsequent case studies, while simultaneously *orientating* them. This is a sensitive requirement, which is difficult to fulfil. As a hypothesis, I will consider general properties inherent to technologies as the point of departure: technologies show regularities and are organized with respect to human purposes in order to be able to fulfill their purposes (following and expanding on Grunwald and Julliard, 2005; Grunwald and Julliard, 2007). Before explaining this approach to debating technology in a general sense, I will briefly give the major arguments in favor of it:

- It seems to be helpful for orientating the subsequent analyses because life in its biological consideration, as we saw above, is also deemed to be organized, namely as living organisms.
- It does not predetermine the following analyses with respect to their results because it refers neither to life nor to nature but rather argues from attributes inherent to technology and engineering, independent of consideration of living entities.
- It is in accordance with a vast majority of images of technology in the philosophy of technology (e.g., Mitcham, 1994), while not substantially fixing, e.g., the role of technology in society or for humankind.

Regularity, reproducibility, and situational invariance have proven to be characteristic of the generalizing discussion on technologies and the concept of technology (Grunwald and Julliard, 2005). They are methodologically centred around technical rules, which form the pragmatic side of epistemic cause/effect chains (Bunge, 1966; 1972). They emphasize the *repeated success* of an action under certain conditions and make the conditions for reproducibility explicit. Technological knowledge represented in technical rules permits the repeated implementation of actions while legitimately expecting that one will always get the same result. Technical rules also allow separation of technical knowledge from a singular context and transferring it to other situations (Bunge, 1972). They make organization, planning, and division of labor possible, by dividing action chains into separate operations such as modules. Technical rules make routines of action available and reliable. They also allow us to transfer knowledge over time, to store it, and to use it for education. The transferability of rules of actions and the reliability they allow are the basis of human cooperation and social organization, also at higher levels of abstraction: certain forms of regularity in situations and stability of processes are prerequisite for the possibility of culture.

Machines are prototypical and even paradigmatic forms of regular organization. They consist of many parts, devices, and materials; they convert energy, organize forces, process data, and have moving elements. They constitute an organized system of many parts, which have to co-function in order to serve the overall functions of the machine (Mitcham, 1994, 161ff.; Nick *et al.*, 2019). These parts have to show high regularity, e.g., a diesel engine or an algorithmic calculator, in order to meet their required purposes in a reliable manner according to human expectations. In the subsequent section, the machine paradigm will be introduced as a possible bridge between *technology* and *life*, allowing us to demonstrate similarities and differences as well as the changing relations between *life* and *technology* (Mahner and Bunge, 1997; Gutmann and Knifka, 2015; Mainzer, 2015; Matern *et al.*, 2016; Boldt, 2018).

Regarding regularities and reproducibility as inherent properties of technology, and thus as generalizable attributes, is shared among established positions in the philosophy of technology (e.g., Mitcham, 1994). Regarding technology as a concept of reflection independent from restrictions to material objects allows us to go a step beyond; regular social relationships, regular decision-making procedures, and rules of social life consolidated by customs and law show a similar structure. Technical rules function analogously to social institutions (Claessens, 1972). The organization of many areas of society follows the model of technical rules for enabling stability and reliability. For example, bureaucracy is an organized form of administrative procedures. Regularization, control, hierarchical structuring of operations, etc. allow bureaucracy to appear to be a "social machine" for creating reproducible results for certain social functions and expectations. The machine character also expresses itself in the fact that a bureaucracy consists of functionally substitutable individual components, with human beings as "cogs in the works" or spare parts of the system. The armed forces can also be interpreted as a technical form of human behaviour. Marching, for example, is a technical manner of locomotion. The technological aspect of communication through the strict hierarchization of the chain of command can be noted as system of technical rules, just as the armed forces as a whole can be interpreted as a technical system. Here, the model of the machine again serves as a bridge-builder at the occasion of technical rules being present in both technology and society.

Although regularity is, therefore, a precondition for social life, and for the stabilization of culture through the ages, it is nonetheless *ambivalent*: The societal ambivalence of *bureaucracy* can easily be recognized by the negative connotations of terms like "bureaucratization." In the case of the military, this ambivalence shows itself whenever one speaks of the militarization of society as a whole. In early 20th century Prussia, technical and military manners pervaded wide areas of society, restricting individual freedom to small areas of life in favor of forming a collective machinery, which then enthusiastically entered the Grand War. These observations demonstrate the deeply rooted *ambivalence of regularization*; on the one hand, securing life requires regularity, on the other, regularity can threaten freedom and individuality. This ambivalence motivates clearly audible questions: "What degree of regularization and reproducibility is necessary, what degree is possible, what is desirable or even merely acceptable, and in which areas?" They pose an issue of particular relevance to concerns of the subordination of humans under technology (Chap. 8).

Readers might wonder why these questions are raised in a book on the changing interfaces between life and technology. As we will see, they play an important role in some of the case studies, at the occasion of a feared technicalization of humans, and a subordination of humans under the realm of technology, in particular in the cases of germline intervention and human enhancement. They also show that technical rules, as the cognitive core of technology, play an important role, not only in biology and at the level of individual human beings but also in society. In a nutshell, we see here a possible overlap in epistemological respects – at the occasion of the machine model – between fields as apparently completely separate as technology, society, and biology, which will be subject to further investigation in the case studies.

2.3   Uncharted Territory: Life and Technology at a Crossroads

The separate introductions of life and technology given above implicitly followed a scheme going back to Aristotle, who strictly distinguished between nature (with life belonging to this realm) and culture (including technology). This dichotomy is deeply rooted in Western culture and philosophy (Cleland, 2019):

> This often-used opposition refers to Aristotle and adopts his distinction between *phýsis* and *téchne* in a modern sense: Living beings are defined as natural objects, which have the source of their growth and movement in themselves. In contrast, artefacts are seen as artificial objects, which are fabricated by human beings and exist only in relation to their use. They belong to culture [...] (Funk *et al.*, 2019, 177).

In many interpretations, the Aristotelian distinction between culture and nature was transferred to the distinction between technology and life (Brenner, 2007). Simultaneously the resulting boundary mostly was taken as an ontological one, strictly separating two different areas of being. This is not the place to investigate whether this appropriately corresponds to Aristotle's writings and intentions (cp., for example, Funk *et al.*, 2019; Gutmann, 2017). My point is only the mighty legacy of this ontological interpretation, which strictly divides the word into an animated and an unanimated part. Taking this separation as a strict and ontological boundary, no bridge would be possible, and phrases such as "engineering life," "fabrication of life," or "life as technical product" (Boldt, 2013) as used, e.g., in synthetic biology (Chap. 4) would not even be understandable. Moreover, they would motivate not only concern but also protest and indignation – as they sometimes really do.

According to the basic observations motivating this book, which will be unfolded in the case studies, we cannot deny that things are changing. Deeper and deeper technical intervention into living entities becomes possible, increasing human influence on them. Consequently, questions arise as to what extent we can still subsume highly manipulated bacteria or animals under the realm of nature in the Aristotelian sense, and whether we are witnessing the emergence of a new dimension of beings somewhere between culture and nature, and between life and technology in the traditional sense (Funk *et al.*, 2019). Then, the established dichotomies between nature and culture or life and technology could become porous and perhaps meaningless, resulting in uncertainty and needs for new orientation. All definitions of life that are based on these dualisms and dichotomies would have to be abandoned (Funk *et al.*, 2019, 177), requiring new orders. The neologism of "biofacts" (Karafyllis,

2008) directly addresses this *new and still uncharted territory* – which is, also accessed from the side of technology through artificial intelligence and autonomous robots (Chap. 8).

If there is new and uncharted territory, then we need conceptual patterns and research strategies for learning and recognition but also for shaping this territory according to our intentions and ethical values. The machine model could serve as a bridge-building notion (Mainzer, 2015; Matern *et al.*, 2016), which was among the results of the previous sections. Biology investigates living entities "as if" they were machines (Gutmann and Knifka, 2015), while engineers are constructing machines for human purposes; both operate with regard to machines as regulated and well-organized entities. The machine model, therefore, will accompany the analyses conducted in this book.

Looking at the "as if" mode of biological recognition (Gutmann and Knifka, 2015; Gutmann, 2017), we can try to describe what has been changing at the boundary between life and technology over recent decades, while still using the dichotomy which is under pressure. If there was, and still is change, then this should be illustratable at the occasion of the "as if" model of biological recognition, e.g., in the field of synthetic biology or animal enhancement. At this level, we can express the following hypotheses of ongoing change, which will be subject to further clarification and scrutiny in the case studies:

- The machine paradigm and related engineering thinking has already become influential and will become more and more powerful (Mainzer, 2015). Biological insight into biotic functions and biotechnical capability for manipulation and creation has grown to a huge extent and will grow further.
- For many scientists, the "as if" mode of operation has become more and more invisible, leading to the perception that living organisms not only could be investigated "as if" they were machines but that they *are* machines. The epistemological "as if" has changed into the ontological "is" in an implicit and creeping manner while confusing the use of models, metaphors, and the modeled entities (Gutmann and Knifka, 2015).
- Human perception of living entities has changed accordingly, beyond science. Humans regard living beings, including

themselves, increasingly as special kinds of machines, composed of different elements and determined by causal laws (Boldt, 2018). Colloquial language is a good indicator for this, e.g., when people talk about their brains and minds by applying computer language.

- Technological advance allows us to empower technology, e.g., robots, with more and more capabilities traditionally owned by living objects. Emulation and simulation of humans and other living entities has become better and better, e.g., humanoid robots. Conversely to the previous point, humans tend to anthropomorphize those technologies (Weber, 2015).

These changes at the formerly (more or less) strict boundary between life and technology lead to a bundle of hopes and expectations, questions and problems, concerns and uncertainties, and a resulting need for orientation and guidance. This book aims at providing some insight as preliminary responses to the following questions:

(1) Referring to the somewhat provocative title of this book: Is something like "living technology" possible, regarding the dichotomy between *life* and *technology* referred to at the top of this section? Is it possible to create life (Boldt, 2013), or living beings, out of its mere non-living elements such as molecules? If yes, is artificially created life different from natural life, and in what respects would it be different?

(2) Referring to the moral status of living objects: What can we say about the moral status of newly created or heavily modified living entities? What is the crucial criterion for assigning moral status, and what kind of moral status depending on which criteria? Is the set of attributes and capabilities, which an object deemed to be "living" is able to perform, decisive, or should the object's ancestry be taken as the deciding criterion? In other words: does the moral status assigned depend on the set of attributes an object shows, or on its biography; whether it is either a descendant of other living entities or was made by engineers? While the latter option corresponds to the Aristotelian view of nature as grown and culture as made, the first-mentioned alternative would open up a completely new field for reflection, shaping, and regulation.

(3) Referring to established ethics: Bio- or pathocentric ethics are well-established fields providing guidance, e.g., for animal welfare. However, in many cases that are becoming at least thinkable, they would no longer suffice. If, for example, animals could be created without the capability to suffer or to feel pain, then pathocentric ethics would run into a dilemma, or become meaningless (Chap. 5). On the other side of the boundary, anthropomorphic views of robots and autonomous systems are spreading. First voices postulate no longer regarding intelligent robots as mere things (Chap. 8). Instead, many postulate moral status should be assigned to them, and machine ethics should regulate their behavior. Formerly clear accountabilities of sub-disciplines of applied ethics no longer work, calling for new approaches.

(4) Referring to responsibility: While the distribution of responsibility has been an issue in the ethics of research and engineering for decades, a new wave of intense debate emerged at the occasion of the disappearing boundaries between life and technology. In the uncharted territory in the midst of the traditional dichotomy between life and technology, new responsibility configurations occur, which must be reflected in order to arrive at new and trustworthy distributions.

(5) Referring to future developments: Regarding the scientific and technological advance of the previous decades, and prolonging it into the future, what can we expect, and what should we fear? Will the future witness the ultimate victory of the machine paradigm and full technicalization of everything? Will this also be the end of humankind, at least the end of humans as we have been developing over millennia?

These five questions and the related issues and concerns offer a formidable set of challenges for analysis, reflection, and shaping future development at the crossroads of *life* and *technology*, while traditional dichotomies lose their importance. The epistemologically nice "as if" mode of biological recognition (Gutmann and Knifka, 2015) could collapse, in the eyes of many people, leaving even us humans as machines, at least in the dominant self-perception of many. Hence, we see huge uncertainties, which by far exceed contemporary technical means and instruments but address visions of the future, images of humanity, new interfaces between humans and technology, and future prospects for society. We are confronted not only with specific questions on regulation, e.g., in cases of germline intervention or animal protection, but also with questions such as what kind of society we *want* to live in, what self-image of humans we implicitly assume, and whether or under what conditions future developments are acceptable, responsible, or desirable.

As a rule, opinions as to what is desirable, responsible, tolerable, or acceptable in society are a matter of controversy, demarcating normative uncertainty and often giving rise to moral conflict. This holds, in particular, for all questions touching upon issues of life related to highly dynamic scientific, technological, and medical advances, which raise new questions and challenges all the time. As soon as these questions lead to normative uncertainties, e.g., with respect to the moral status of embryos or robots, then ethical analysis and reflection is needed to provide orientation and guidance.

Chapter 3

Ethics Guiding Our Pathways to the Future

The new and envisioned technical and scientific opportunities to transgress the boundary between *technology* and *life* allow for innovation in many areas and have huge transformative potential. To realize this potential responsibly, to exploit its benefits with regard to human interests, and to handle possible unintended side effects in a careful and precautionary manner needs *orientation*. In particular, it is the breathtaking novelties provided by research fields such as human genome editing, enhancement, robotics, and synthetic biology that create open questions and normative uncertainty because living entities are affected. Therefore, ethical reflection, complemented by philosophical inquiry, is needed to support opinion forming in societies and political decision-making for funding and support, but also adequate regulation.

Preparing the search for orientation in the subsequent case studies needs some preliminary clarifications regarding the understanding of ethics, in particular with respect to different approaches to ethics and the variety of expectations with respect to ethical analysis and reflection, e.g., between Continental and Anglo-American understandings of ethics or between analytical and normative ethics. The introduction of ethics for guiding the development of new configurations between *life* and *technology* will

Living Technology: Philosophy and Ethics at the Crossroads Between Life and Technology
Armin Grunwald
Copyright © 2021 Jenny Stanford Publishing Pte. Ltd.
ISBN 978-981-4877-70-1 (Hardcover), 978-1-003-14711-4 (eBook)
www.jennystanford.com

begin with a brief sketch of ethics in general (Sec. 3.1), followed by a consideration of ethics in and for science and technology (Sec. 3.2), and by a focused introduction of the key notion of "responsibility" and of the ethics of responsibility (Sec. 3.3).

3.1 Problem-Oriented Ethics

In the understanding to be applied in this analysis, ethics is a reflective discipline aimed at providing moral orientation to fields of human practice. Therefore, ethics is dedicated to contribute to *problem-solving* by providing *orientation* (Sec. 3.1.1). This point of departure implies the necessity to answer the question about the specific problems and challenges, as well as their origins, which ethics should or could address. The conceptual answer will be given by distinguishing between *standard* (Sec. 3.1.2) and *non-standard* situations, in a moral respect (Sec. 3.1.3).

3.1.1 Ethics for Analyzing and Resolving Moral Conflicts

The distinction between factual morals guiding practical action and decision-making and ethics as the reflective discipline for considering morals, in particular in the case of conflicts or ambiguities, has been widely accepted (e.g., Habermas, 1992). Morals govern social practice: "[m]oral argument and exploration go on only within a world shaped by our deepest moral responses" (Taylor, 1989, 5). They can be researched empirically as part of the "is," while ethical reflection needs normative argumentation and patterns of justification for dealing with the "ought." While morals are directly action-guiding, ethics serves the reflexive and deliberative resolution of conflict situations, which result from the actions or plans of actors, based on inconsistent, insufficient or divergent moral conceptions.

The aim of ethical reflection is to support or even allow the continuation of action and decision-making in cases of different or diverging moral conceptions in an argument-based and peaceful manner. In other words: ethical reflection is required if the normative guidance of the morals factually observed in the respective context does not suffice to orientate the respective action or decision. This holds for all situations where, for a given action or decision-making problem, there is no consensually accepted moral background from which orientation for decision-making can be derived, or where the morals in force do not suffice to clearly orientate the respective situation. In the following, situations involving lack of clarity at the level of action-guiding morals will be denoted as situations of *normative uncertainty*.

One implication of determining ethics as a reflective activity to contribute to overcoming normative uncertainty by providing new orientation is simply that ethical reflection is not required in the absence of normative uncertainty. This conclusion affects all areas of human life: ethical reflection is not required in many, probably even in most of the cases when humans act, or face the need to make a decision. Indeed, looking at human life-worlds and daily practices, it is clear that actors do not reflect from an ethical perspective on all of the actions and decisions performed; on the contrary, ethical reflection is undertaken rather rarely, while most of the decisions taken are parts of daily routines. They follow established action schemes under clear normative orientation and are, thus, below the threshold for ethics. This view corresponds to the approach of Habermas (1973), who distinguished between the levels of action (*Handlungsebene*) and discourse (*Diskursebene*). The latter is only addressed at the occasion of considerable problems and challenges, which cannot be resolved at the level of action by established routines, action schemes, and moral orientation. The rapidly developing fields at the crossroads between life and technology provide a vast expanse of missing orientation, which is at the core of the motivation for writing this book. Ethical reflection then should provide orientation by analyzing as well as reflecting and transcending the previously existing moral background.

Ethical reflection can thus be seen as a methodological approach which applies methods, procedures, instruments, and tools to discursively analyze and resolve conflicts arising from different moral assumptions in situations of normative uncertainty. Ethics in this sense of searching for normative orientation in those situations consequently always deals either with questions of collective actions or of individual actions with a collective impact and is, therefore, closely related to political philosophy (see below). The distinction between ethics and morals takes into account the plurality of morals in modern society, which gives individuals wide autonomy to determine and practice their own ideas of a good life. It restricts ethics to those cases where pluralism and individual freedom meet limits because the rules and procedures of *collective* life must be determined in order to ensure peace, fairness, and justice for all members. Individual, group, or community ideas of the good life and of a good society, representing the respective moral conceptions and preferences, have to be assessed, weighed, and compared to investigate whether there are conflicts between particular morals and how these could be overcome in order to allow continuation of social practices with action and decision-making. Following the conceptualization of the relation between politics, the public, and democracy proposed by American philosopher John Dewey,

> [...] democracy is the regulation of the public interest arising from indirect consequences and related conflicting interests; it is combined with the idea that everyone should be involved and, in principle, regarded as a person capable of co-deciding about a regulation of such indirect consequences (Dewey, 1927, 147, quoted after Kowarsch, 2016, 20f.)

we consider ethics as argument-based advice to democratic reasoning, deliberation, and decision-making.

The task of ethics is thus to analyze morals with respect to the argumentation patterns and their normative grounds, their presuppositions and premises, their normative content and implications, and their justifiability with respect to *argumentative rationality*. In the history of philosophy some key approaches for the principles of reflection have been developed, such as the Virtue Ethics by Aristotle, the Categorical Imperative proposed by Kant, and the idea of the Pursuit of Happiness formulated by Bentham (Utilitarianism). In this volume, Habermas' discourse ethics (1992), which originates in an analysis of implications of communicative rationality, will be used as umbrella theory. The universal and obligatory nature of morality is, according to Habermas, grounded in the universal rationality of communicative action.

Scrutinizing the validity of moral prescriptions and values can, according to this theory, be performed analogously to the justification of facts, namely in discourses between proponents and opponents observing characteristic rules of fairness and inclusion. As an example, critical hermeneutics offers opportunities to look "behind actions, namely the hidden ideological-moral views, focusing on their practical consequences in our practices at the societal level" (Roberge, 2011). Argumentative rationality can be methodically achieved by organizing the entire argumentation pro and con in a transparent series of steps, e.g., in the form of "if–then" chains (Grunwald, 2003). Mutual understandability of all the steps in moral argumentation and of the statements made among the participants of the discourse includes understanding the conditions of validity of the knowledge and of the justifiability of norms and values proposed. Following the postulate for transparency by Dewey (1931), Jürgen Habermas (1970) postulates that the advice provided by experts to the public and policymakers should be *transparent*. In this way, ethical reflection, just like other forms of scientific policy advice, must be open to the public, in contrast to the models of a decisionistic or a technocratic relationship between science, the public, and politics. Accordingly, ethical policy advice can only play an advocatory and preparatory role for decision-making.

This positioning of ethics includes multiple relationships between ethical reflection on the one hand, and social practice, culture, and the life-world on the other. It implies that, in spite of the principal ethical universalism of discourse ethics, or of Kantian and utilitarian principles, applied ethics must be performed in close relation with real-world actors. It must include a careful view on its various relationships with social practices, and so is not separated from society in an ivory tower but is part of society and its ongoing processes of communication, deliberation, and action. Strong contextualization is needed, e.g., in fields such as the role of technology in care homes, technology at the end of human life, and human enhancement politics. Consequently, problem-oriented ethics needs appropriate knowledge about the affected field of society and its governance in order to be able to engage the appropriate addressees.

3.1.2   Standard Situations in a Moral Respect

We will now look more closely at the issue of *normative uncertainty* by differentiating between standard and nonstandard situations, in a moral respect. Among the basic observations motivating this book (Chap. 1) was the thesis that new developments at the interface between life and technology create a lot of moral uncertainties, which must be addressed in order to be able to harvest the benefits expected. A field repeatedly addressed throughout this book is the question of the moral status of newly created entities belonging to the spheres of both life and technology.

In social practices, explicit ethical reflection hardly ever forms our initial observation. Most decisions take the form of goal – means deliberations at the action level (Habermas, 1973), without any particular reflection on their normative background, for example, in everyday routine actions. The discourse level, from where this background could be called into question, is the rare exception (see above). It is practically neither possible nor sensible to subject the entire arsenal of normative contributions in society to continuous ethical reflection, which would immediately result in the standstill of all social activities. In this sense, the great majority of decisions can be classified as "standard" in the following sense (this argumentation goes back to Grunwald, 2000, which was developed further in several papers): the normative basis for the decision is not made the object of special reflection but accepted *as given* in the respective situation. Ethical reflection is, for example, not required when engineers in a lab think over the question of whether they should use iron or aluminum for a certain component, or when a licensing authority has to decide on an application for building a biotechnological manufacturing plant. In all of these cases, normative elements such as criteria for weighing alternative options obviously play an important role – however, without any need for ethical reflection, because the criteria to be applied already give clear orientation for the decision-making processes and are in accordance with other moral frameworks, which have to be observed. These could be national and international legal regulations, planning protocols, agreements on good practice, the standard procedures of the relevant institutions, e.g., corporate guidelines on the supply chain, possibly the code of ethical guidelines of the profession concerned, as well as observed but uncodified social and cultural customs.

The question then poses itself of how to distinguish between situations requiring ethical reflection because of normative uncertainty and situations without that necessity. My proposal is to categorize situations along the distinction of standard and non-standard, in a moral respect, and to make this distinction operable by offering the following set of criteria that must be fulfilled for assigning a particular situation the "standard" attribute (expanding on Grunwald, 2000):

(1) *Pragmatic Completeness*: The normative framework must pragmatically provide sufficient coverage of the pending decision from a normative point of view. None of the essential aspects can be left unanswered or considered to be indifferent from a normative point of view. This criterion cannot be satisfied in situations for which routine attitudes are not yet present but contested (Sec. 3.1.3). Well-known examples are tied to advances in science and technology, such as atomic weapons, genetic diagnostics, the use of animals for experiments (Chap. 5), or the implementation of autonomous technology in human–technology interactions (Chap. 8).

(2) *Local Consistency*: There must be a sufficient degree of consistency among the elements of the locally relevant normative framework in the respective field of action. "Sufficient" means that the conclusions for the pending decision or opinion converge, even if they follow from different morals (e.g., different religious attitudes). Otherwise, there would be a conflict leading to normative uncertainty, for which ethical reflection would become necessary. In the case of stem cell research, for example, different attitudes as to whether an embryo is to be attributed the status of a person lead to conflicting judgments regarding the circumstances under which such research can be carried out (Chap. 6).

(3) *Joint Understanding*: Sufficient consensus regarding the interpretation of the normative framework governing a situation must exist among the actors in the context of the pending decision. It is useless, for example, if all of the involved persons or positions proceed from "the dignity of man" but understand this differently. Here, normative uncertainty would occur, e.g., for human genome editing (Chap. 6).

(4) *Acceptance*: The normative framework has to be accepted as the basis of a decision by the actors and those affected. Where this is not the case, there is a conflict in a normative sense, and thus normative uncertainty, which would render it impossible to speak of a moral standard situation. This would also be the

case if there were no agreement about the selection of the criteria to be employed to choose among different options for acting. An example is whether cost–utility considerations or ethical criteria based on animal welfare considerations should be employed (Chap. 5).

(5) *Compliance*: The normative framework not only has to be accepted in the respective field, but it must be *de facto* adhered to by the majority. When this is not the case, a conflict between valid regulation as part of the normative framework and factual behavior would occur. Such a situation (cp. the debates on abortion in many countries) is nonstandard, even though ethical reflection will hardly be able to correct the consequences.

If all of these conditions are satisfied in a specific context, then normative uncertainties do not exist in preparing for certain action or decision-making. Consequently, there is no need for ethical reflection. Participants and others affected by a decision can take information about the normative framework into consideration as axiological information without having to reflect, because the currently valid normative frameworks – according to the criteria listed above – provide undisputed and comprehensive orientation for acting and decision-making.

3.1.3   Beyond Standard Situations in a Moral Respect

Social coherence, including its morals and normative frameworks, is incomplete and unstable, a fragile mesh which is continually being challenged and threatened by social developments or by innovations and the conflicts resulting from them. Morals and normative frameworks are rooted in social practice and can therefore change over time: a once-valid normative framework can be put in doubt by even minor changes in certain parameters. Since transgression of the limits of an acknowledged normative framework can change standard into non-standard situations, in a moral respect, the challenge consists in being able to recognize these shifts at all. This observation leads to the postulate for careful and continuous observation and monitoring.

Scientific results and technical innovations (but not only these) often challenge existing normative frameworks by raising new questions while making ethical reflection necessary (cp. several examples in this book, e.g., synthetic biology and human enhancement). New scientific knowledge and technological innovation, but also social, legal, or political novelties as well as a change of values and lifestyle, may transform earlier standard situations, in a moral respect, into *nonstandard situations* where one or more of the criteria given above are no longer fulfilled. In this modified situation, there are three options to choose from:

(1) *Reject the innovation* causing moral trouble, with the implication of also renouncing its possible benefits, in order to maintain the initial normative framework. As a rule, this option must be chosen if there are strong, i.e., categorical, ethical arguments against the new technology. We will discover some examples in the subsequent case studies, e.g., for human genome editing (Chap. 6).

(2) Modify the properties of the innovation causing moral trouble, maybe the circumstances of its production involving animal experiments, or the location of a nuclear waste disposal site in a region sacred to indigenous people, in order to be able to harvest the expected benefits without causing moral trouble. To put it briefly: make the innovation compatible with the framework. The focus is on designing the innovation along the ethical factors so that the innovation fits the framework. Examples might be taking measures for increasing biosafety and biosecurity for synthetic biology (Chap. 4), or introducing care robots in accordance with the values of people affected, rather than in a top-down approach.

(3) Modify the normative framework, so that the innovation could be accepted and the benefits harvested in a way that would not lead to problems regarding the modified framework. Examples are animal experiments undertaken for nonmedical purposes (see Chap. 5), or research in which the moral status of embryos plays a role (Chap. 6). The issue is then to examine if and to what extent the affected normative framework, e.g., the national legislation, could be modified in accordance with

basic principles of the constitution and without coming into conflict with essential ethical principles.

Ethical reflection plays a decisive role for anticipated actions in determining the choice between these alternatives and possible further differentiations. In approaches 2 and 3, the reflection is an act of balancing the expected advantages of the innovation or the new technology against the moral or other costs if – as in probably the most common situation – there are no *categorical* ethical arguments for or against. Approach 1 is different in this respect because it presupposes that balancing would be ethically not legitimate, e.g., if "human dignity" would be violated, according to the German constitution.

All these reflections and decisions have to be made in specific configurations of human actors, regulations, traditional and existing morals, cultural backgrounds, and different political governance regimes. Ethical reflection plays a different role in each of these configurations, according to the actors and governance regimes, which differentiate the pragmatic constellations for ethics in general and ethics for science and technology in particular. Without any claim to completeness, these include:

- debates in research policy on priorities in funding research projects and programs;
- debates about regulations of biotechnology and the life sciences;
- discussions about the responsibility of scientists and engineers or other actors in the context of new biotechnology;
- discussions in medicine about new diagnostic and therapeutic procedures with consequences even for the doctor–patient relationship or for the cost structure in the healthcare system;
- controversial issues in dealing with biomaterials and the related processes in research, manufacturing, and production;
- debates in the media and among the public, frequently triggered by civil society organizations and social movements;
- debates among different groups of the population as well as between differently developed regions of the world about the distribution of the advantages expected from new biotechnology;

- debates about societal futures in view of the far-reaching promises made for new biotechnology.

Ethical reflection has, therefore, to play a role in many different areas also far beyond academic philosophy, including different groups of actors and giving advice to various fields of action and decision-making. This observation will guide the discussion of responsibilities of actors involved and people affected in the subsequent case studies. Ethics is not separated from ongoing and perhaps heavy debate but part of it. Among the largest challenges might be to get an audience for argument-based ethical reasoning, even in highly controversial public debates.

3.2 Ethics for Guiding the Techno-Scientific Advance

This brief section is intended to serve as a bridge between the more general picture of ethics drawn above and the following chapters on specific developments at the interface between *life* and *technology*.

It has long been a matter of controversy whether science and engineering have any morally relevant content at all. Until the 1990s, technology was frequently held to be *value neutral*. Numerous case studies have, however, since recognized the normative and value background of the development and use of technology and made it a subject of reflection (e.g., van de Poel, 2009). Basically, technology does not consist just of a growing set of artefacts and processes but rather is deeply embedded in societal processes (Rip *et al.*, 1995). Technology does not originate on its own and does not develop according to inherent regularities but is consciously produced by human actors rooted in interests and values in order to reach certain ends and purposes. Technology is, therefore, embedded in individual or societal goals, problem diagnoses, and action strategies from its initial stages. There is no pure technology in the sense of a technology completely independent of this societal dimension (Radder, 2010). Technology is human-made, therefore morally relevant, particularly concerning its purposes and goals, the measures and instruments used, and the evolving effects and consequences. Technology is currently recognized as being an appropriate object for moral responsibility (Jonas, 1979; Durbin, 1987) and ethical reflection (van de Poel, 2009).

This is also true of science. The value neutrality of science was postulated in the era of positivism. Since then, there have been many developments clearly demonstrating the necessity of considering ethical aspects of science as part of human responsibility. Examples are the atomic bomb, which generated a far-ranging debate on the responsibility of physicists; the genetic modification of organisms, which motivated consideration of issues of responsibility as early as the conference of Asilomar in 1975; the cloned sheep Dolly in 1997; or the birth of twins in China after intervention into their germline in 2018 (Chap. 6). In many fields, science – analogously to technology – does not operate by distantly contemplating how nature works; it is instead involved in societal purposes and strategies and not only explains nature but also delivers knowledge for action, manipulation, and intervention. The notion of technoscience reflects on the disappearance of the strict traditional distinction between pure science and applied technology in many fields of investigation (Asdal *et al.*, 2007), in particular in the so-called new and emerging sciences and technologies (NEST) (Swierstra and Rip, 2007).

Regarding ethics as a reflective discipline for *de facto* moral issues means that ethics will achieve practical relevance for science and technology only once normative uncertainties arise. According to the very nature of science and new technology, they provide society with *innovation*, i.e., with new knowledge and new opportunities for technical intervention into nature as well as into society. Therefore, it is not an exception but the rule that new technology will raise new questions for its responsible implementation and use, leading quickly to spaces beyond standard situations, in a moral respect: "The global challenges of terrorism, immigration, international politics, and most importantly, technoscience, need a compass, which is missing today" (Ruggiu, 2019). For example, the question of responsible use of prenatal or pre-implantation diagnostics and therapy did not make any sense as long as the respective technologies were not available. The availability of new technology often motivates new, sometimes completely new questions. As soon as these questions lead to normative uncertainties (e.g., with respect to intervention into the identity of humans by germline intervention), then ethical analysis and reflection is required (Hansson, 2017).

Such uncertainties often exceed technical means and instruments but are about visions of the future, about images of humanity and future prospects for society, e.g., addressing the future of the relation between society and nature, or between humans and technology. In this way, scientific, political, and public debates about new technologies often find themselves approaching questions such as what kind of society we *want* to live in, what self-image of humanity we implicitly assume, and whether or under what conditions future developments are acceptable or desirable. As a rule, opinions as to what is desirable, tolerable, or acceptable in society are a matter of controversy, demarcating normative uncertainty and often giving rise to moral conflict. Conflicts over technology are, as a rule, not only conflicts over the means of technology, e.g., in questions of efficiency, but also conflicts over visions of the future, concepts of humanity, "dignity" of life, animal welfare, and views of a just society. Therefore, it does not seem exaggerated to view science and technology as not only one field of ethical reasoning required amongst many others but rather as a field of eminent significance for determining the further course of human development.

The morally relevant aspects which science and technology exhibit can be distinguished into three categories according to a simple action-theoretical structure, concerning (1) the purposes they pursue, (2) the instruments they employ, and (3) the consequences and side effects they produce (cp. Hansson, 2017).

(1) Agenda-setting in technology and science depends on epistemic but also on practical aims and goals, e.g., what science and technology should contribute to meeting future challenges. In many cases, these are not problematic and remain in standard situations, in a moral respect. To develop therapies for illnesses such as Alzheimer's disease, to provide new facilities to support disabled people or to protect society against natural hazards – visions of this type can be sure to gain high social acceptance and ethical support. In other areas, however, there are social conflicts even at the level of aims and goals. Some visions related to crewed spaceflight, the vision of improving human performance, or visions of giving political control to algorithms are examples of controversial subjects, in a moral respect. These questions lead to the challenges of

what knowledge and which technologies we want to have in some future and what we do not want to have.

(2) The instruments, measures, and practices applied in research and development may lead to moral conflicts and normative uncertainty. Examples are the moral legitimacy of experiments using animals (Chap. 5); practices making use of human persons, embryos, or stem cells as subjects of research; the use of patient data for scientific research; experiments with genetically modified organisms or plants, especially their release beyond laboratories or, in earlier times, the experimental testing of nuclear weapons. Questions regarding the moral status of human embryos and of animals, as well as questions regarding the acceptability of the risks involved in research have led to heavy moral conflicts.

(3) Since the 1960s, the unintended and adverse effects of scientific and technical innovations have increasingly been perceived as a serious problem. Severe accidents occurred in technical facilities (Seveso, 1976; Bhopal, 1984; Chernobyl, 1986; Fukushima, 2011). Threats to the natural environment occurred at the global level (air and water pollution, ozone hole, climate change, loss of biodiversity). Negative health effects, as in the case of asbestos, and problematic social effects (e.g., labor market problems caused by automation) also have clearly demonstrated the deep *ambivalence* of technology. The technological advance not only provides new opportunities for improving health, welfare, security, safety, and sustainable development but also has a "dark side." Unfortunately, as philosopher Hans Jonas (1979) stated, the positive consequences of technology usually are deeply and inseparably intertwined with their negative side effects (Sec. 2.2). His diagnosis was that the most significant ethical aspects of technology do not result in malfunctioning technologies but rather in smoothly *functioning* technology. Negative effects often accompany the intentional use of technology in a slowly emerging and creeping manner, which is difficult to discover. Examples are the ozone hole and climate change, which evolved over decades before their discovery. Both are the consequence of functioning technologies – not of malfunctioning or dysfunctional ones. The same holds for

problematic consequences of the use of Social Media and the Internet for democratic deliberation. This situation leads to societal and ethical challenges: How can a society that places its hopes and trust in innovation and progress protect itself from undesirable, possibly disastrous side effects, and how can it preventatively stockpile knowledge to cope with possible future adverse effects? What extent of risk or ignorance is morally acceptable? How is responsible action possible in view of the high uncertainties involved? Questions of this kind demarcate deep normative uncertainties behind, e.g., the role of the Precautionary Principle (e.g., Persson, 2017; Sunstein, 2005; von Schomberg, 2005), or other attempts at providing orientation (Grunwald, 2008a).

In all of its activities, ethics serves as *advice* to science, engineering, stakeholders, research funding, regulation, public debate, etc., not as a kind of authority. Ethical reflection on new science and technology in the three directions mentioned needs close cooperation of professional ethics and philosophy with scientists and researchers deeply involved in research and development such as biology, medicine, robotics, and neuroscience. Ethical debate must not be a discourse external to ongoing scientific and engineering research but should rather be considered an interdisciplinary part of it, complemented by inclusive measures of integrating citizens and stakeholders when required.

3.3   Ethics of Responsibility

Responsibility has been a key concept in the ethics of technology since its early days (Jonas, 1979). In this respect, the previous decade witnessed the emergence of the concept of responsible research and innovation (RRI, Owen *et al.*, 2013; van den Hoven *et al.*, 2014; von Schomberg/Hankins, 2019; Pellé and Reber, 2015). Building on these traditions, responsibility will be used as an umbrella concept integrating ethical, empirical, and epistemic issues (Sec. 3.3.1). Because the latter shows itself to be crucial regarding the subject of this book; the shifting boundaries between *technology* and *life*, consequentialist ethics of technology (Sec. 3.3.2), and ethics beyond consequentialism (Sec. 3.3.3) will be distinguished.

3.3.1 The Understanding of Responsibility

We usually talk about responsibility if there is a reason, i.e., if issues such as the distribution of responsibility or its range into the future are controversial. This can, for example, be the case if new opportunities for action are made possible by new technology for which there are still no rules or criteria for the attribution of responsibility available, or if the latter are a matter of controversy. Unclear or controversial assignments of responsibility can be the reason for changing a standard to nonstandard, in a moral respect. The purpose of speaking about responsibility is then to overcome these normative uncertainties and achieve agreement over the structure of responsibility to be applied in the affected field, e.g., in fields such as synthetic biology (Chap. 4) or autonomous technology (Chap. 8). Speaking about responsibility thus ultimately serves a practical purpose: clarification of the specific responsibilities for actions and decisions. "Responsibility ascriptions are normally meant to have practical consequences" (Stahl *et al.*, 2013, 200). The significant role of the concept of responsibility in discussions of scientific and technological progress and for dealing with its consequences is obvious (Lenk, 1992). Providing orientation by clear and transparent assignment of responsibility, however, needs a common understanding of responsibility.

Usually, a more or less clear meaning of the notion of responsibility is assumed for applying it in everyday communication. However, this supposition mostly is not fulfilled in more complex fields, e.g., in responsibility debates on future science and technology. Here, a more in-depth scrutiny of the concept of responsibility is required (Grunwald, 2014b, building on Lenk, 1992; cp. also Fahlquist, 2017). Responsibility is the result of social processes, namely of *assignment acts*, whether actors take responsibility themselves, or the assignment of responsibility is made by others. The assignment of responsibility follows social rules based on ethical, cultural, and legal considerations and customs (Jonas, 1979, p. 173). They take place in concrete social and political spaces involving and affecting concrete actors in concrete constellations. Accordingly, a five-place reconstruction for discussing issues of responsibility in scientific and technical progress will be applied in this book (Grunwald, 2014b):

- *someone* (an actor) assumes responsibility or is made responsible (responsibility is assigned to them) for
- *something* (such as the results of actions or decisions, e.g., on the R&D agenda in a specific field or on risk management)
- *before an instance* (a social entity expecting particular responsibilities from its member[s] which perhaps transfers duties – this may be a religious community, the entire society, the family, etc.) with respect to
- *rules and criteria* (in general the normative framework governing the respective situation, e.g., rules of responsible behaviour given in a Code of Conduct), and relative to the
- *knowledge available* (knowledge about the impacts and consequences of the action or decision under consideration).

The first two places are grammatically trivial in order to make linguistic sense of the word "responsible." Semantically the first three places indicate the fundamental social context of assigning responsibility, which inevitably is a process among social actors and thus constitutes the *empirical dimension* of responsibility, as described above. The fourth place opens up the *ethical* dimension of responsibility (also described above), while the fifth place addresses an additional dimension by referring to the knowledge available about the object of responsibility (in place two). It forms the *epistemic dimension* of responsibility. Consequently, the resulting "EEE model of responsibility" comprises:

- The *empirical dimension* of responsibility takes seriously that the assignment of responsibility is an act by specific actors, which affects others, and mirrors the basic social constellation of assignment. Attribution of responsibilities must take into account the ability of actors to influence actions and decisions in the respective field, regarding also issues of accountability, power, and legitimation. Relevant questions for responsibility analyses are: How are the capabilities, influence, and power to act and decide distributed in the field considered? Which social groups, including scientists, engineers, managers, citizens, and stakeholders, are affected and could or should help deliberate and decide about the distribution of responsibility? Should the questions under consideration be debated at the *polis* or can they be delegated to particular groups? What consequences would a particular distribution of responsibility have for the governance of the respective field, and would it be in favor of the desired developments?

- The *ethical dimension* of responsibility (cp. also Grinbaum and Groves, 2013; Pellé and Reber, 2015; Gianni, 2016; Ruggiu, 2018) is reached when the question is posed regarding the *criteria and rules* for judging actions and decisions under consideration as responsible or irresponsible, or for helping to find out how actions and decisions could be designed to be (more) responsible. Relevant questions about responsibility reflections are: What criteria distinguish between responsible and irresponsible actions and decisions? Is there consensus or controversy on these criteria among the relevant actors? Which meaning of normative issues such as "dignity of life" or "animal welfare" is presupposed, and by whom? Can the actions and decisions in question (e.g., about the scientific agenda or about containment measures to prevent risks) be regarded as responsible with respect to the rules and criteria?

- The *epistemic dimension* asks about the knowledge of the subject of responsibility and its epistemological status and quality. This is relevant in particular in debates on scientific responsibility because frequently statements about impacts and consequences of science and new technology show a high degree of uncertainty. The comment that nothing else comes from "mere possibility arguments" (Hansson, 2006) is an indication that in debates over responsibility it is essential that the status of the available knowledge about the futures to be accounted for is determined and critically reflected from an epistemological point of view (Gannon, 2003; Nordmann, 2007a; Grunwald, 2014a). Relevant questions in this respect are: What is really known about prospective subjects of responsibility? What could be known in the case of more research, and which uncertainties are pertinent? How can different uncertainties be qualified and compared to each other? And what is at stake if worse comes to worst?

Debates over responsibility in technology and science often focus on the *ethical dimension*, while considering issues of assignment processes and epistemic constraints to be secondary issues. However, regarding the analysis given so far the ethical dimension is important but only part of the picture. It might be that the familiar criticisms toward responsibility reflections (see above) of being simply appellative, of epistemological blindness, and of being politically naïve are related to narrowing responsibility to its ethical dimension. Instead, relevant questions in responsibility debates arise in *all* of these three dimensions, which, therefore, must be considered together in prospective debates in our field (following Grunwald, 2014b). In particular, answers to the question of what could be sensible objects of responsibility (position two of the introduction of responsibility above) regarding future developments at the life–technology interface strongly depend on the epistemological dimension, as will be demonstrated in the following.

3.3.2   Consequentialist Ethics of Technology

Since the origins of the ethics of technology, unintended and undesired consequences of technology have been at the center of its reflection (Jonas, 1979). In particular, technology-induced risks and adverse long-term effects, e.g., for the natural environment, have been focuses of ethical reflection over recent decades. Normative uncertainties emerged from the use of new technology because formerly unknown effects, such as the possible risks of synthetic nanoparticles, climate change, and massive loss of biodiversity occurred as consequences, motivating questions of regulation, change of behavior or other measures of dealing with them responsibly. The desire to earn the benefits of new technology has repeatedly led to conflicts with desires to keep the natural environment in a sustainable status. Issues of responsibility emerged to overcome these conflicts and inconsistencies between social practice and normative desires (Jonas, 1979; Lenk, 1992) and have continued until today without resolution. The *ambivalence* of technology (see above), resulting not only in desired effects but simultaneously in adverse effects, is the origin of this development. Therefore, the ethics of science and technology is to a large extent *consequential ethics* (Darwall, 2002).

Consequential ethics focuses on the expected consequences and impacts of actions and decisions rather than on the intentions standing behind them. Even in a Kantian approach – which looks primarily to the *intentions* of action rather than on expected consequences – it would be unavoidable to take into account assumed consequences in order to get a comprehensive picture of the situation under consideration. In a certain sense, ethical reflection of what should be done, or not done, is closely related to considering the consequences of actions and decisions, independent of the specific ethical approach. In science and technology, this is quite obvious conceding that intentions should be considered, too. Ethical judgment of new science and technology primarily has to look at intended but also at possible non-intended consequences (Fig. 3.1).

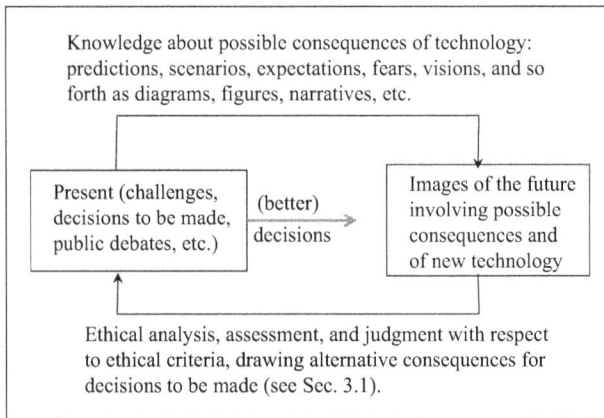

Figure 3.1 Consequentialist ethics providing orientation for the present by deliberating and assessing future consequences (Source: Grunwald, 2012, 313, modified)

The notion of consequentialism, however, must not be used in too narrow a sense. Often, the ethical assessment of consequences of action is combined with a utilitarian approach. Ethical analysis is thus transformed into balancing risks and benefits using a quantitative calculus, often in monetary terms. By aggregating these data, the utility of the considered options of decision-making can be calculated and the option showing the highest expectable utility can be identified, as in cost–benefit analysis. However, consequentialist reasoning can also be conducted by applying other approaches to ethics, such as deontological, Aristotelian, or Rawlsian approaches, operating in a more qualitative and argument-based manner.

Consequentialist ethics comprises the utilitarian approach but is not bound to that specific type of reasoning.

In fact, the utilitarian approach mostly is not applicable at all if new and emerging technologies are to be addressed with little knowledge about any consequences being available. Independently from the ethics approach taken, reflections of consequentialist ethics of technology are usually accompanied by challenges due to the uncertainty of knowledge and the question of the ability to bear responsibility in this situation (Hansson, 2010). As a prominent example, the Precautionary Principle included in the environmental regulation of the European Union (von Schomberg, 2005) reflects on how to deal with possible adverse affects to the environment in the absence of valid knowledge:

> Where, following an assessment of available scientific information, there is reasonable concern for the possibility of adverse effects but scientific uncertainty persists, measures based on the precautionary principle may be adopted, pending further scientific information for a more comprehensive risk assessment, without having to wait until the reality and seriousness of those adverse effects become fully apparent (quoted after von Schomberg, 2005).

The Precautionary Principle can be understood as an attempt to make taking responsibility possible, even when the epistemological dimension within the EEE approach (see above) renders doing this almost impossible. However, the controversial discussion around this principle shows that there is still need for improvement and reflection, in particular on the meaning of the crucial notion of a "reasonable concern" required for any application, instead of a mere suspicion (cp. Harremoes *et al.*, 2002; Peterson, 2007; Weckert and Moor, 2007).

The Precautionary Principle was implemented following the experience that traditional approaches to introducing new technology such as "wait and see" or "trial and error" are not responsible if possibly far-ranging and irreversible adverse consequences could occur. It was the debate on GMOs in the 1990s that motivated introducing the Precautionary Principle (von Schomberg, 2005) and that was again taken up at the occasion of the debate on synthetic nanoparticles (Weckert and Moor, 2007; Grunwald, 2008a).

A crucial challenge to the ethics of science and technology in general is behind the debates around this principle and the problems in making it operable, which applies in particular to new and emerging science and technology (Swierstra and Rip, 2007; Grunwald, 2008a; Fahlquist *et al.*, 2015). It is the question for the adequate relation in time between the scientific-technological advance, on the one hand, and ethical reflection and technology assessment, on the other (Decker and Fleischer, 2010). Transferring the Control Dilemma (Collingridge, 1980) to this field, concerns have been expressed of ethics coming either too late to be able to influence anything, or coming too early, without having valid knowledge of the consequences of new technology to reflect upon (Mnyusiwalla *et al.*, 2003). Ethical reflection in the "ethics first" approach is exactly confronted with high uncertainty or even ignorance about the consequences of specific advances in science and technology.

This observation motivates the question of the epistemological preconditions, which have to be fulfilled to make the consequentialist approach possible (Grunwald, 2013). As can be seen from the model (Fig. 3.1), the availability of prospective knowledge about consequences is crucial. Its creation by foresight or forecasting is not an end in itself: the prospective knowledge is only the object of assessment, reflection, evaluation, and judgment. If the anticipated consequences made object to ethical scrutiny were merely speculations, all the conclusions drawn would be arbitrary. Therefore, the epistemic quality of the prospective knowledge is the backbone of consequentialist reasoning.

3.3.3 Responsibility Beyond Consequentialism

Two established modes of creating orientation in the consequentialist paradigm are using *prognostic futures* and *thinking in scenarios* (Grunwald, 2013). The prognostic mode presupposes the availability of knowledge about regularities or laws, be it causal or statistical; this knowledge legitimates, under certain conditions, extrapolating past developments, e.g., trends, to the future, allowing use of the resulting predictions as orientation for action. Thinking in scenarios applies a more open approach by considering a broader range of possible but *plausible* futures. In order to exploit scenarios developed, e.g., in the field of future energy supply, the diversity of the scenarios must be limited. Often this happens by limiting the future developments regarded as plausible between a worst and a best scenario. It is essential that the boundary between "plausible" and "implausible" futures is clear, transparent, and accepted. Otherwise, drawing any conclusions for action would necessarily fail (Grunwald, 2013). This repeatedly happened and still happens in typical NEST fields (new and emerging science and technology) dominated by visionary communication, often wavering between apocalyptic fear and paradise-like expectation (Grunwald, 2007). The following quote taken from a visionary paper of synthetic biology (Chap. 4) allows serious doubt about the possibility of responsibility debates in such cases at all:

> Fifty years from now, synthetic biology will be as pervasive and transformative as is electronics today. And as with that technology, the applications and impacts are impossible to predict in the field's nascent stages. Nevertheless, the decisions we make now will have enormous impact on the shape of this future (Ilulissat Statement, 2007, 2).

This statement expresses that (a) the authors expect synthetic biology will lead to deep-ranging and revolutionary changes, (b) today's decisions will have high impact on future development, but (c) we do not know what those future impacts will look like. If this were true there would not be any chance to assign responsibilities, and even to speak about responsibility would be without purpose because there would not be any valid subject to talk about (similar to the "mere possibility arguments" problem, cp. Hansson, 2006). This constellation holds true for many debates on future science and technology, in particular to those related to futures of the life–technology interfaces.

> However, this way of understanding responsibility tends to assume a consequentialist perspective that cannot answer to the uncertainty that characterizes the development of innovative techniques and technologies. RRI's crucial issue, the one for which we make use of the criterion of responsibility, is exactly to provide an answer to the uncertainties that are implied in the complex relations between individual actions, social relations, and natural events (Gianni, 2016).

Therefore, if neither the prognostic nor the scenario-based modes of orientation work, a deep shift of perspective is needed as to how orientation could be provided in the epistemologically disastrous situation where the consequentialist loop of reasoning (Fig. 3.1) breaks down (Grunwald, 2014a). In this situation, debates are widely conducted by using visions, utopias, dystopias, and other more narrative stories of the future (Sand, 2018). While these future statements usually cannot be classified in epistemological respect, they nevertheless can have far-ranging consequences:

> The importance of visions is that they have a performative dimension. In other words, these representations of the future tend to modify the present in accordance with our expectations of the future (Ruggiu, 2019).

This factual power of speculative narratives and visions cannot be tackled in any of the approaches mentioned (Grunwald, 2017). The space of debated futures is too broad while the futures frequently are highly contested (Brown *et al.*, 2000). These concerns and challenges have been formulated based on observations in fields such as synthetic biology (Chap. 4), human enhancement (Chap. 7), and autonomous technology (Chap. 8). The divergence of many communicated NEST futures together with the impossibility to distinguish "plausible" from "implausible" futures in epistemic respect seem to destroy any hope of gaining orientation by reflecting on future developments. Therefore, we have to seek for other ways of realizing responsibility.

Responding to this challenge, the *hermeneutic* mode of orientation was proposed (Grunwald, 2013). In situations of extremely high uncertainty, or complete openness of the future, we only can investigate and understand the various and diverging future stories, including their background, and explore *what these say about us today*. The divergent futures can be examined for what they mean and under which diagnoses and values they originated. We cannot learn anything out of this about how the future plausibly will become. But in the hermeneutic mode, we can learn something for and about our present situation (Grunwald, 2014a).

The basic observation goes back to Augustine of Hippo (397). Futures in general and techno-futures in particular do not exist *per se* but rather are human-made and constructed in a more or less complex manner. Techno-futures, whether they are forecasts, scenarios, plans, programs, or speculative fears or expectations, are produced using a whole range of ingredients such as the available knowledge, perspectives shared in a research community, value judgments, and *ad hoc* assumptions. They have authors with specific purposes and intentions. All of the elements used to build images of the future are, thus, elements of *present* time. No data of the future are available.

Therefore, the hermeneutic analysis of pictures of the future hardly tells us anything about the future in the sense of the time to come but rather about *issues today*. If future statements, e.g., about care robots, human enhancement, or germline interventions, are interpreted in this way, then we can learn something *explicitly* about ourselves, our societal practices, subliminal concerns, implicit hopes and fears, and their cultural roots by investigating all the diverse stories of the future *as expressions of today*. The primary issue is then to clarify the meaning, origin, and biography of the speculative futures: What is at issue; which rights might possibly be compromised; what images of humankind, nature, and technology are formed, and how do they change; which anthropological issues are involved; what perspectives are taken concerning human relations to animals; and what designs for society are implied in the projects for the future? What remains once any claim to anticipation is abandoned is the opportunity to view the lively and controversial debates about NEST and other fields of science or technology not as anticipatory, prophetic, or quasi-prognostic talk of the future but as expressions of present time.

Reminding ourselves of the fact that all these visionary and speculative futures have been created by authors with intentions and motives, that these futures constitute interventions into the real world, and that they can have more or less far-reaching consequences, be they intended or unintended, it becomes easy to address responsibility issues beyond a traditional consequentialist paradigm (Simakova and Coenen, 2013; Grunwald, 2017). Creation and dissemination of technology futures can be interpreted in terms of their responsibility, just as any action can. However, the object of responsibility here is not the distant future and the possible consequences of today's NEST developments, as in consequential ethics, but their creation and dissemination *as such* in present time.

The ethical aspect concerning responsibility does not emerge from the fact that the debated futures might become reality but rather merely from the fact that they disseminate pictures of the future today with an interventional power (Grunwald, 2017; Sand, 2018).

By summarizing this issue, we can state that hermeneutic orientation corresponds to an epistemologically different perspective, while considering stories about the future as expressions of contemporary observations, diagnoses, perceptions, hopes, and fears. Telling stories about futures belongs to a respective present time. At least sometimes, those stories adhere to some *Zeitgeist* convictions, which could become outdated. Looking back from an imagined year 2040 could, in principle, render various issues of today bound to our time today, e.g., the fears of losing control over AI, the fear of running into totalitarian systems because of full surveillance, or the fear of compromising the "dignity of life" while crossing boundaries between life and technology. Unfortunately, this imagined look back can only be imagined based on knowledge, data, and convictions of today. In epistemological respect, we are bound to the "immanence of the present" (Grunwald, 2007) while needing considerations of futures for orientation purposes. There is no way to escape from this situation in the absence of valid knowledge.

Chapter 4

On the Track to Creating Life: Synthetic Biology

Over recent decades, scientific research and biotechnological capabilities have made deeper and deeper interventions into living organisms possible. Huge advances in nanobiotechnology and systems biology, which increasingly benefit from the potential of ongoing digitalization, have fueled the emergence of synthetic biology and raised high expectations that humans may perhaps be able to create living systems from scratch in some not too far distant future. After more than 200 years of synthetic chemistry and about twenty years of synthetic nanoparticles, a new wave of human "creation" is ongoing. The question of whether and how such developments can be performed in a responsible manner has been raised in an intensified form since about the late 2000s, building on earlier reflections on the ethical issues raised by biotechnologies and genetically modified organisms (GMO).

This chapter reviews the state of the art of the responsibility debate regarding synthetic biology and extends this by applying the framework of problem-oriented ethics and responsibility (Chap. 3). First, a brief introduction of synthetic biology and its roots in nanobiotechnology will be given (Sec. 4.1). Second, the societal and ethical debate, including considerations of opportunities and risks, will be reviewed, based on the results of projects on ethical,

Living Technology: Philosophy and Ethics at the Crossroads Between Life and Technology
Armin Grunwald

Copyright © 2021 Jenny Stanford Publishing Pte. Ltd.
ISBN 978-981-4877-70-1 (Hardcover), 978-1-003-14711-4 (eBook)
www.jennystanford.com

legal, and social aspects (ELSA) and on technology assessment (TA) studies (Sec. 4.2). The third part will systematically elaborate on the ethical issues, which demarcate normative uncertainty at the occasion of focal items of ethics, such as the moral status of created organisms (Sec. 4.3). The responsibility configuration of synthetic biology through the lens of the EEE model developed in the preceding chapter will be analyzed in Sec. 4.4. Finally, issues beyond consequentialist ethics will be addressed: Sec. 4.5 includes a hermeneutic consideration of the changing relationship between technology, nature, and life, considering implications and repercussions for the notion and understanding of life itself.

4.1   Synthetic Biology

Synthetic biology turned into a vibrant field of scientific inquiry and ethical debate around the year 2005 (Giese *et al.*, 2014). In 2010, Craig Venter, one of the pioneers of synthetic biology, announced that he had successfully implanted artificial DNA into a bacterium. Synthetic biology then rapidly became known to the public (Synth-Ethics, 2011) and has led to a series of activities on its ethical, legal, and social implications (ELSI; cp. Sec. 4.2) as well as to responsibility reflections and analyses (Sec. 4.4).

The combination of engineering at an extremely small scale, e.g., by nanotechnology (see below), and biology, in particular molecular biology and systems biology, is at the roots of synthetic biology. Since the diameter of DNA, a typical object of technical operation, is approximately two nanometers (nm), synthetic biology can be considered "a specific discipline of nanobiotechnology" (de Vriend, 2006, 23; see below). In turn, synthetic biology can also be viewed as the continuation of molecular biology using the means of nanobiotechnology. In the meantime, synthetic biology has also become a sub-discipline of biology. While nanotechnology involves the development of materials and machines at the nanoscale, synthetic biology builds on the insight that nature already employs components and methods for constructing something similar to machines and materials at very small scales. By employing off-the-shelf parts and methods already used in biology and by developing new tools and methods, synthetic biologists hope to develop a set of tools to accelerate deepening human influence on living systems by new technology (Synth-Ethics, 2011).

Synthetic biology differentiates between an approach that uses *artificial* arrangements of molecules to (re)produce biotic systems and one that combines elements of classical biology to form new systems that function in a manner beyond pre-existing nature (Benner and Sismour, 2005). The motivation behind this is to create artificial or technically modified forms of life that are partially equipped with new functions. The major question of synthetic biology is: "how far can it [life] be reshaped to accommodate unfamiliar materials, circumstances, and tasks?" (Ball, 2005, R3). Examples range from the design of artificial proteins, to the creation of virus imitations or the reprogramming of viruses, and even extend to attempts to program cells to perform desired functions (Ball, 2005; Benner and Sismour, 2005, 534–540).

Various definitions have been suggested for synthetic biology, all of which point in the same direction despite different accentuations. Accordingly, synthetic biology is seen as (cp. also Pade *et al.*, 2014, for an in-depth consideration of several novelties of synthetic biology):

- The design and construction of biological parts, devices, and systems and the redesign of existing, natural biological systems for useful purposes (LBNL, 2006).
- The design and synthesis of artificial genes and complete biological systems and the modification of existing organisms, aimed at acquiring useful functions (COGEM, 2006).
- The engineering of biological components and systems that do not exist in nature and the re-engineering of existing biological elements; this is determined by the intentional design of artificial biological systems, rather than by the understanding of natural biology (Synbiology, 2005).
- The use of a mixture of physical engineering and genetic engineering to create new (and therefore synthetic) life forms (Hunter, 2013).
- Applying the engineering paradigm of systems design to biological systems in order to produce predictable and robust systems with novel functionalities that do not exist in nature (EC, 2016).

All these definitions touch upon a common concept: the creation of new biological systems via the synthesis or assembly of artificial and natural components. To do this, synthetic biology needs to encompass a broad range of methodologies from various disciplines, such as genetic engineering, molecular biology, systems biology, membrane science, biophysics, chemical and biological engineering, electrical and computer engineering, control engineering, and evolutionary biology ("Synthetic biology." *Wikipedia*. https://en.wikipedia.org/wiki/Synthetic_biology; accessed 20 May 2020).

The epistemic approach of synthetic biology regards biotic units (living organisms) as complex technical relationships, which can be broken down into simpler technical ones. This approach could be named "deconstructing life" following the model of technology (de Vriend, 2006). Living systems are examined within the context of their technical function, and cells are interpreted as machines – consisting of components, analogous to the components of a machine, which have to co-operate in order to fulfill the overall function. For example, proteins and messenger molecules are understood as such components that can be duplicated, altered, or newly compounded in synthetic biology. A modularization of life is thereby made, as well as an attempt to identify and standardize the individual components of life processes (Danchin, 2014). In the tradition of technical standardization, gene sequences are saved as models for various cellular components of machines. While this is still, so to speak, a form of *analytic* biology, it becomes *synthetic* as soon as the knowledge about individual processes of life obtained from technical modeling is combined with the corresponding experiments and utilized so that certain useful functions can be achieved (Schwille, 2011). Following design principles of mechanical and electrical engineering, the components of living systems should be put together according to a building plan in order to obtain a functioning whole. The recombination of different standardized bio-modules (sometimes called "bio-bricks") allows for the design and creation of different living systems. Artificial cells based on such components and micro-machines are projected that can, for example, process information, manufacture nanomaterials, or make medical diagnoses. Entirely in the tradition of mechanical and electric engineering, such machines are supposed to be built part by part according to a design that is drafted top-down in order to enable useful purposes: "Seen from the perspective of synthetic biology, nature is a blank space to be filled with whatever we wish" (Boldt and Müller, 2008, 388). With the growing collection of modules, out of which engineering can develop new ideas for products and systems, the number of possibilities grows exponentially. The result is supposed to be a functioning entity:

> Engineers believe it will be possible to design biological components and complex biological systems in a similar fashion to the design of chips, transistors, and electronic circuits (de Vriend, 2006, 18).

The *combination* of knowledge about molecular biology and genetic techniques with the opportunities offered by nanotechnology is decisive for scientific and technological progress. The prerequisite for a precise design of artificial cells would be a sufficiently thorough understanding of all the necessary subcellular processes and interactions. In this respect, synthetic biology is still at an early stage of development. Many research fields contribute to improving the state of the art, with nanobiotechnology being among the major predecessors (Schmid *et al.*, 2006).

The concept of nanobiotechnology, or bionanotechnology, (cp. Jotterand, 2008a) was coined in the context of the National Nanotechnology Initiative of the United States (NNI, 1999). Its point of departure is the observation that basic life processes take place at a nanoscale, which is the size of life's essential building blocks, such as proteins. At this level, nanotechnology could make it possible to engineer cells by networking natural biological processes with technical ones. Visions of nanomachines at the level of cellular and subcellular processes take the form of mechanisms for producing energy, molecular factories, and transport systems or high-capacity data storage and data reader systems. The language of engineering is extended to processes and objects of life at the nanolevel:

> To a large extent, the metaphors in synthetic biology are borrowed from the fields of engineering, construction and architecture, electrotechnics, information theory or information technologies (IT), computer science, design and theology (Funk *et al.*, 2019, 179).

Examples of such uses of language refer to hemoglobin as a "vehicle," to adenosine triphosphate synthase as a "generator," to nucleosomes as "digital data storage units," to polymerase as a "copier," and to membranes as "electrical fences." Functional biomolecules act as components for gathering and transforming light, or as signal converters, catalysts, pumps, or motors:

> The fact that biological processes are in a way dependent on molecular machines and clearly defined structures shows that building new nano-machines is physically possible (Kralj and Pavelic, 2003, 1011; cp. also Danchin, 2014).

A characteristic example is the attempt to create technical replicas of photosynthesis (Cheng and Fleming, 2009; acatech, 2018). Plants and some forms of bacteria assure their energy supply by means of photosynthesis (Blankenship, 2014). Sunlight is used to synthesize complex carbohydrates from carbon dioxide and water, which serve both for energy storage and as energy supply. In contrast to current photovoltaic cell technology, this principle even functions in diffuse or very weak light. The idea of using the principle of photosynthesis, as it has developed in the course of evolution, to technically ensure an energy supply for humans is exceptionally appealing (acatech, 2018). Energy supplied on the basis of this principle would be CO_2 neutral, would be easily storable, could be produced in a decentralized fashion, would be practically inexhaustible, and would not produce any problematic waste (Faunce *et al.*, 2013). Nanobiotechnology provides the techniques needed to understand the natural processes at the molecular level and possibly to be able to replicate them. It concentrates on replicating the simpler manner of functioning in bacteriochlorophyls, relying on the principle of *self-organization* for the formation of the corresponding nanoscale structures (Balaban and Buth, 2005). There is hope that such research can contribute to the development of engineered biosensors and artificial antennas that can even function in weak and diffuse light (French *et al.*, 2014). They could then be useful for the design of hybrid solar cells based on economical polymer technologies (Balaban and Buth, 2005, 207). Such research is however still entirely in the sphere of basic research (acatech, 2018; cp. the references provided there). The point is to understand essential processes in the context of their technical functioning (Erb and Zarycki, 2016). Designations such as "light harvesting complex" or "proton pump" demonstrate the technical view of photosynthesis on processes stemming from life (Cheng and Fleming, 2009).

Approaches for utilizing the principles of evolution to achieve certain new effects are another approach in synthetic biology. For example, cells could be subject to the pressure of artificial evolution by turning off the genetic sequences responsible for the building of certain amino acids. By adding chemical substances that are chemically sufficiently similar to the missing amino acids, the cell can be brought to use the substitutes in place of the amino acids. The result of this is a cell with modified properties. Here there is a tight interface with systems biology (Bruggeman and Westerhoff, 2006; Boogerd *et al.*, 2007), in which the complex interaction of the many individual processes is to be understood as a complex entity:

> Systems biology is crucial to synthetic biology. It includes knowledge about the natural basic biological functions of RNA and DNA sequences in information storage, energy supply, membrane functions, cell structure, cell-to-cell signalling, gene regulation (gene expression), and metabolic functions in natural systems [...] (de Vriend, 2006, 23).

The traditional self-understanding of biology, which is molded by the natural sciences, aims to *understand* vital processes. In epistemic respect, a widespread understanding among synthetic biologists is that understanding life, as is the traditional aim of biology, will be fully achieved only if biology becomes able to *rebuild life*. Because rebuilding existing life and inventing new types of life requires the same type of knowledge, the aim of understanding leads to the capability of creating life by synthetic biology (Ball, 2005; Woese, 2004). Synthetic biology as an engineering science is about a *new invention* of nature and the creation of artificial life on the basis of knowledge about traditional and "natural" life. It is no longer satisfied with investigating life which already exists but aims at redesigning or even reinventing nature. Early successes have been reported (cases taken from *Wikipedia*):

- A completely synthetic bacterial chromosome was produced in 2010 by the team of Craig Venter and introduced to genomically emptied bacterial host cells (Gibson *et al.*, 2010).

- In 2019, researchers reported the creation of a new synthetic (possibly artificial) form of simplified but viable life (Fredens *et al.*, 2019).

This change of perspective from understanding to creating life transforms biology into a technical science (de Vriend, 2006) that embodies the dual strands of cognition and design and that is subordinate to the primacy of design goals. This even holds at the current stage of development, with only a few applications. Just as in the classical technical sciences, synthetic biology is more "know how" than "know that" or "know why" (Pade *et al.*, 2014). Though the latter are both required for providing "know how," they only keep an *instrumental* function in the research processes of synthetic biology:

> Although it can be argued that synthetic biology is nothing more than a logical extension of the reductionist approach that dominated biology during the second half of the twentieth century, the use of engineering language, and the practical approach of creating standardised cells and components like in an electrical circuitry suggests a paradigm shift. Biology is no longer considered "nature at work", but becomes an engineering discipline (de Vriend, 2006, 26).

Therefore, synthetic biology can be subsumed under the concept of *technosciences* (Latour, 1987; Asdal *et al.*, 2007), in which the area of life is modelled as an ensemble of machines (Danchin, 2014). In this new form of biology, "the pre-existing nanoscale devices and structures of the cell can be adapted to suit technological goals" (Ball, 2005, R1). This position has consequences for the applicability of the consequentialist paradigm of the ethics of technology (Sec. 3.3.2) and assignments of responsibility (see Sec. 4.4).

4.2 The Debate on the Opportunities and Risks of Synthetic Biology

Research and development on any new technology is accompanied by classifying envisaged opportunities as well as feared risks (Giese and von Gleich, 2014; König *et al.*, 2016). These debates usually motivate questions of responsibility – responsibility for harvesting the opportunities while minimizing the risks. Therefore, we will first take a view of the landscape of opportunities and risks of synthetic biology, which has been emerging since about 2006.

Although synthetic biology was almost completely unknown outside small circles of experts until approximately 2006, a public discussion has developed rapidly since that time, in particular following some spectacular news disseminated by Craig Venter (Synth-Ethics, 2011). Research agencies quickly reacted and funded several projects in the fields of technology assessment (e.g., de Vriend, 2006; Sauter *et al.*, 2015) as well as on ethical, legal, and social aspects (ELSA) of synthetic biology. Most of the issues relevant today were identified at that early stage, e.g.:

> It [synthetic biology] has potential benefits, such as the development of low-cost drugs or the production of chemicals and energy by engineered bacteria. But it also carries risks: manufactured bioweapons and dangerous organisms could be created, deliberatively or by accident (Church, 2005, 423).

The second World Conference on Synthetic Biology in 2006 brought about the first interest among CSOs (civil society organizations) (ETC Group, 2007). In view of the fact that, compared to traditional gene technology, synthetic biology further increases the depth of human intervention into living systems (Pade *et al.*, 2014), discussions on precautionary measures (Paslack *et al.*, 2012; von Gleich, 2020) and on the responsibility of scientists and researchers emerged. Their first manifestations were provided in the form of several ELSA and TA activities (Sec. 4.2.1). Based on these studies, an overview of the current perception of the opportunities (Sec. 4.2.2) and risks (Sec. 4.2.3) will be given.

4.2.1 Studies on the Opportunities and Risks of Synthetic Biology

Several ELSA and some TA studies in this field have been performed since 2006. Funding agencies and political bodies such as advisory commissions soon recognized the importance of insights into opportunity *versus* risk weighting, possible ethical challenges, and possible conflict situations with the public. Some examples are (following the presentation in Grunwald, 2016a):

Ethical and regulatory challenges raised by synthetic biology – Synth-Ethics

Synth-Ethics, funded by the European Commission, was among the first ELSA projects on synthetic biology, approved in 2007. It applied a special focus on biosafety and biosecurity and explored challenges for the notions of life. It also analyzed early public debates around these issues and identified challenges for current regulatory and ethical frameworks. Finally, it formulated policy recommendations targeted at the synthetic biology community, EU policymakers, NGOs, and the public (see https://www.itas.kit.edu/english/projects_grun09_synth-ethic.php, accessed 28 Sep. 2020).

Engineering life

Funded by the German ministry for education and research (BMBF), this project had the following objectives: (1) to investigate whether synthetic biology would enable humans to create life and what this would mean in ethical respects; (2) to analyze the rhetoric phrase of "playing God" from a philosophical and theological perspective; (3) to explore risks and opportunities of synthetic biology in a comprehensive manner; and (4) to scrutinize legal boundary conditions for research in synthetic biology (see University of Freiburg, n.d.).

Synthetic Biology

This project was commissioned by the German *Bundestag* and conducted by its Office of Technology Assessment (TAB, cp. Grunwald, 2006). In addition to the scientific-technological aspects, the main issues were: ethics, safety, and security, intellectual property rights, regulation (or governance), public perception, the issue of synthetic biology as a "do-it-yourself" technology, and adequate and early communication about opportunities and risks (Sauter *et al.*, 2015).

SYNENERGENE – Synthetic Biology: Engaging with New and Emerging Science and Technology in Responsible Governance of the Science and Society Relationship

The aim of the EU-funded SYNENERGENE project was to initiate activities with a view to stimulating and fostering debate on the opportunities and risks of synthetic biology. Among other issues, it monitored developments in synthetic biology, identified critical aspects, experimented with diverse participation formats – from citizen consultations to theatrical debates – and engaged stakeholders from science, the arts, industry, politics, civil society, and other fields in the debate about synthetic biology (see https://www.itas.kit.edu/projekte_coen13_senergene.php; accessed 28 Sep. 2020).

Presidential Commission

The Presidential Commission for the Study of Bioethical Issues (2010) advising the U.S. President explored potential benefits of synthetic biology, including the development of vaccines and new drugs and the production of biofuels that could someday reduce the need for fossil fuels. It also addressed the risks possibly posed by synthetic biology, including the inadvertent release of a laboratory-created organism into nature, and the potential adverse effects of such a release on ecosystems. The Commission urged the policy level to enhance coordination and transparency, to continuously perform risk analysis, to encourage public engagement, and to establish ethics education for researchers.

This brief overview of some ELSA and TA activities allows for concurrent conclusions concerning more general and overarching issues:

- While the focus of the considered activities varies according to the respective setting, the issues addressed show considerable overlap. Some issues such as biosafety and biosecurity appear in all of the studies. They have established a kind of "canon" of ELSA issues of synthetic biology.
- Understanding the novelty of synthetic biology, of its promises and challenges compared to previous approaches, is a significant part of all the studies (cp. also Pade *et al.*, 2014).
- The set of values and the normative background applied in the assessments vary from case to case. They are diverse, partially controversial and contested, indicating the presence of some normative uncertainty (Sec. 4.3).
- Lack of prospective knowledge about future innovation paths and products emerging from synthetic biology as well as about

the possible consequences of their development and use was reported in all of the studies, thereby pointing to limits to the consequentialist approach of ethics (Sec. 3.3.3).

Summarizing these results of studies, which were mostly performed at the early stage of synthetic biology, it is striking that in spite of high uncertainty and little knowledge about the consequences of synthetic biology at that time, a concurrent view was reached on the major issues to be monitored in the years to come. This result approves the postulate for the early involvement of TA and ethics, even in cases of little knowledge.

4.2.2   Overview of Opportunities

The aim of synthetic biology is to design biological systems for numerous useful purposes, such as for the production of alternative fuels or new materials (e.g., biodegradable plastics). Artificial or redesigned organisms might also be used in medicine, for example, as biosensors, or to help degrade environmental pollutants. The almost complete absence of restrictions limiting possible applications means there is immense room for thinking about potential applications. Synthetic biology could thus provide a considerable contribution to solving many urgent health and environmental problems.

Many applications of synthetic biology are in particular expected in medicine, the life sciences, and biotechnology. In medicine, examples that deserve to be emphasized are new diagnostic means (such as biosensors and the possibility of permanently monitoring the status of someone's health), developments in drug delivery (e.g., the procedure for precisely transporting active substances and targeting their deposition), and the use of new biocompatible materials and surfaces (e.g., Freitas, 2003). The following areas of application can be distinguished in synthetic biology:

- Nanomanufacturing and nanostructuring using bio-based methods. The goal is to use the principle of the self-organization of molecular units to create more complex structures in order to achieve certain functions.
- Engineering of natural proteins and creating novel protein structures that match or improve on the functionality of existing proteins.

- Production of desired chemicals by means of bacteria that are manufactured or modified in a dedicated manner (plan drug factories; de Vriend, 2006, 30).
- The technical utilization of functional biomolecules and of hybrid systems (e.g., of biomolecular motors and actuators) in technical systems or in combination with non-biological components.
- The realization of functions at the interface between biological and technical materials and systems (e.g., for neuro-implants or prostheses).

These areas of application and the corresponding functional combinations make it possible for biosensors and biomembranes (French *et al.*, 2014), for instance, to be employed in environmental technology, and for photoenergetic processes to provide biological support for photovoltaics. For example, it would be economically and ecologically significant if synthetic biology were technically able to copy photosynthesis (see above). Scenarios for similarly far-reaching applications can also be found in biomedicine. Furthermore, the techniques developed by synthetic biology could be used in the production of biofuels (acatech, 2018). Solving the energy problem in a sustainable manner is one of the great challenges facing today's and tomorrow's energy policy and technology development. The creation of synthetic organisms that could help produce biofuels in an efficient way which would not lead to conflicts with the food supply is an attractive idea to many people.

The far-reaching hopes placed in the technical use of processes of self-organization, independent of the concrete fields of application, constitute a point that deserves a great deal of attention. The utilization of phenomena of self-organization, including the possibility of replicating "living things" that have been created or modified by technology, is a central aspect of synthetic biology:

> The paradigm of *complex, self-organizing systems* envisioned by von Neumann is stepping ahead at an accelerated pace, both in science and in technology [...]. We are taking more and more control of living materials and their capacity for self-organization and we use them to perform mechanical functions (Dupuy and Grinbaum, 2004, 292).

There are also economic reasons for the fascination exerted by self-organization. If we could succeed in making materials organize themselves in such a manner that technically desired functions or properties would result, this might be substantially less expensive than if we had to specifically create them, such as by means of a nanotechnological top-down procedure in which atoms are given a specific arrangement. The goal would thus be to substitute construction processes that are designed laboriously by manual action with processes that run by themselves. Human intervention would thus be moved to a different level. The builder would become the controller of a nature that carries out the construction:

> The self-assembling properties of biological systems, such as DNA molecules, can be used to control the organization of species such as carbon nanotubes, which may ultimately lead to the ability to "grow" parts of an integrated circuit rather than having to rely upon expensive top-down techniques (de Vriend, 2006, 19).

Summarizing, it is quite clear that synthetic biology provides a huge amount of potential in very different fields of application. Hence, synthetic biology can be regarded as an *enabling technology*, being key to a heterogeneous range of innovation pathways, similar to nanotechnology in its early stage around the turn of the millennium (Fleischer *et al.*, 2005). This promising configuration is however simultaneously a challenge with respect to ethics and responsibility assignments (Sec. 4.4).

4.2.3 Categorization of Risks

The primary fears of risks connected with synthetic biology concern possible unintended negative consequences for our health and the environment (Giese and von Gleich, 2014) but also the intended utilization of this technological potential for novel biological weapons:

> In the case of synthetic biology, specific risks in need of close scrutiny and monitoring are uncontrolled self-replication and spreading of organisms outside the lab, and deliberate misuse by terrorist groups or individuals or by "biodesigner-hackers" (Boldt and Müller, 2008, 387).

These threats stand for two general areas, namely, safety and security. In our context, the former refers to harms that might occur unintentionally, while the latter concerns harms that occur intentionally. While biosafety is about preventing the unintentional exposure to pathogens and toxins, or their accidental release, biosecurity aims to prevent the unauthorized access, loss, theft, misuse, diversion or intentional release of biological materials.

Let us first look at biosafety. A well-known scenario from the debate about organisms modified by genetic engineering (e.g., Breckling and Schmidt, 2014) concerns a possible product of synthetic biology (e.g., an artificial or modified virus) that might escape from a laboratory and cause considerable risks to our health or the environment without there being a way of recapturing it. Risks of this type involve biological safety (de Vriend, 2006). Beyond posing an immediate danger to living species and individuals, the genetic pool of certain species might be contaminated as a result of genetic transfer, which would lead to ongoing and irreversible modifications.

A major cause of such expectations is that synthetic biology could create living things that are alien to the natural biosphere and for which we do not possess any evolutionary experience for dealing with in ecosystems and organisms (according to de Vriend, 2006; and analogously to the case of synthetic nanoparticles, Grunwald, 2008a). The consequences of the release of such partially or completely invented living things might thus be impossible to anticipate and to calculate. One (accurate) response to this is that the probability is very small that such artificial living things could survive and cause damage in the natural world because they are not adapted to natural processes. But even a small probability is a finite one, which is why problems of this nature require careful observation. A "green goo" was described as the scenario of such a catastrophe, in analogy to the well-known "grey goo" thought by some to be a possible consequence of our losing control of self-replicating nanorobots (Joy, 2000; cp. also Boldt, 2014, 237):

> The green goo scenario [...] suggests that a DNA-based artificial organism might escape from the lab and cause enormous environmental damage (Edwards, 2006, 202f).

This scenario shows the ambivalence of referring to the potential of self-organization. Although self-organization has been used above to refer to capacities that are considered positive and whose technical use promises many advantages, especially economic ones, the capacities of living systems for self-organization and for self-replication also pose a special type of potential threat. As the debate about the gray goo scenario in nanotechnology showed, positive visions can quickly turn into negative ones (Grunwald, 2007). The technical utilization and organization of living processes lead us to a fundamental and inherent risk; namely that we can lose control of living systems in a much more dramatic manner than we can of technical systems because they pass on their properties and can multiply (cp. Sec. 4.3.1).

Following the debate about the risks of GMOs (cp. Breckling and Schmidt, 2014), the Precautionary Principle was developed and implemented in European law for dealing with challenges posed to biosafety (see von Schomberg, 2005; Persson, 2017; Sunstein, 2005; Peterson, 2007). Within the framework of a gradual step-by-step procedure and with a commitment to carefully examine at every step the consequences of that step, a layered containment strategy was established. Accordingly, research initially takes place under high-security conditions, subsequently in standard laboratories, then in controlled open experiments, and finally, currently on application, in controlled plantings that keep a minimum distance from fields used for non-genetically modified agriculture. This has led to a successive reduction in our initial complete ignorance about the consequences of GMOs in our natural environment (Paslack *et al.*, 2012).

The essence of the biosafety argument is that synthetic biology, by intentionally creating new or technically modified living objects, intervenes in the course of natural evolution to a considerable and possibly uncontrollable degree. While natural evolution only takes place in small steps and large modifications only take place over extremely long periods of time, humans are now endeavoring to take control of evolution in just a few years or decades by creating artificial life:

> Ponder for a moment the incredible hubris of the entire endeavor of bionanotechnology. The natural environment has taken billions of years to perfect the machinery running our bodies and the bodies of all

other living things. And in a single generation we usurp this knowledge and press it to our own use (Goodsell, 2004, 309).

In this sense, the massive *acceleration* of natural development by means of synthetic biology procedures is a special challenge, which can be viewed as a risk factor. Artificial cells, even if they are based on knowledge gained from natural cells, will possibly only have a few years of experiments in a laboratory behind them, not millions of years of evolution. This concern mirrors attitudes toward familiar genetic engineering (Breckling and Schmidt, 2014). The new construction of cells or the reprogramming of viruses must, therefore, be placed under special observation and containment.

Finally, the complexity of the processes in molecular biology per se also poses a gateway for potential risks, as has already been shown in the initial experiments with forms of gene therapy:

> A more concrete rationalization of this distrust is the feeling that unpredictable consequences can follow from rearranging a complex system that is not fully understood. The recent history of gene therapy offers us a cautionary tale. To introduce genes into patients, it has been necessary to use what is essentially a piece of nanotechnology – an adapted virus is used to introduce the new genetic material into the cell. But fatalities have occurred due to unexpected interactions of the viral vector with the cell (Jones, 2004, 214 ff.).

The biosecurity argument focuses on the possibility of the *intentional* construction of novel biological weapons on the basis of newly constructed or modified cells (de Vriend, 2006, 54). The products or techniques of synthetic biology, or perhaps only the knowledge produced by it, might be misused for military purposes in government weapons programs or by terrorists. Our fantasies have more or less free rein with regard to the concrete possibilities, and in view of the fact that hardly anything is known about military programs of this nature, there is a danger of chasing after conspiracy theories. One consequence of this state of ignorance is that purely imagined details shouldn't be made widely known as a prediction, such as those regarding intentionally reprogrammed viruses. On the contrary, the issue should be the ethical aspects of the *possibility* of such developments, e.g., in the context of the Precautionary Principle (Persson, 2017). In this respect, it should be taken into consideration that high-technology techniques must be applied in order to create synthetic biological weapons:

> Contrary to popular belief, however, a biological weapon is not merely an infectious agent but a complex system consisting of (1) a supply of pathogen [...]; (2) a complex formulation of chemical additives that is mixed to stabilize it and preserve its infectivity and virulence during storage; (3) a container to store and transport the formulated agent and (4) an efficient dispersal mechanism (Tucker and Zilinskas, 2006, 39).

This means it is no easy matter to produce and use such biological weapons. Since the scientific and logistic effort required would be considerable, such a development is rather improbable among terrorists. Due to this observation, the risk of synthetic biology being performed as a "do-it-yourself" technology seems not that dramatic (Sauter *et al.*, 2015).

Nonetheless, scenarios have been created even for this risk. Synthetic biological weapons could be created by, for example, a fanatic lone operator, an expert in synthetic biology who could misuse their specialist knowledge to develop such living objects out of personal hate and use them against the object of their hate. Or a biohacker, analogous to computer hackers, could construct a damaging virus just to demonstrate that it was technically possible, or to rouse public awareness (de Vriend, 2006). Also in some countries the military has almost limitless resources at its disposal and hardly any logistic problems. In this sense, and despite all the uncertainty and improbability involved, the conscious creation of harmful synthetic or technically modified living objects must indeed be considered a potential risk of synthetic biology. The situation gets even more difficult in face of the dual use dilemma, which

> [...] arises in the context of research in the biological and other sciences as a consequence of the fact that one and the same piece of scientific research sometimes has the potential to be used for harm as well as for good (Miller and Selgelid, 2006).

Of special relevance to synthetic biology are the dual use concerns raised by the *publication* of research relating to biological agents with a large capacity for harm. Key examples given in the literature are the publication of the genomes of engineered virulent mousepox virus, polio virus (Selgelid, 2007), and the 1918 influenza virus (Sharp, 2005). The dual use dilemma means that:

> [...] publishing studies like these [mousepox and polio] both alerts potential bioterrorists to possible biological weapons and actually gives them explicit instructions to produce them [...] [while] the mousepox and polio studies may advance medical science and biodefense preparation (Selgelid, 2007).

All these arguments show that the risks associated with synthetic biology are unclear risks (Grunwald, 2008b). They cannot be tackled in the paradigm of established risk management, e.g., by determining thresholds: neither quantitative data on the probability and extent of damage nor clear knowledge about cause–effect relationships is available. This situation creates challenges to the established consequentialist approaches based on quantitative methods, e.g., cost–benefit analysis and multi-criteria decision analysis.

4.3   Ethical Challenges

Research processes in synthetic biology and possible results such as products usable, e.g., in medicine are not value-free (Hagen *et al.*, 2016). As we saw in the previous section, value-laden issues of safety, interventions into living systems, and self-organization come into the frame. Therefore, the assertion that

> The technology of Synthetic Biology provides a new set of tools. Any ethical challenges come from the way we use the tools and not from the tools themselves (Chopra and Kamma, 2006)

is unacceptable under all circumstances, especially since such assertions of value neutrality cannot even be accepted in the traditional sphere of engineering, which does not touch upon life (e.g., van de Poel, 2009; Radder, 2010). In this section, we look for *specific* challenges that synthetic biology poses to ethical reflection (Gutmann, 2011a). In the programmatic framework of this volume, this means that particular challenges at the interface between technology and life should be addressed.

Indeed, several normative frameworks are affected by synthetic biology (Gutmann, 2011a; Hagen *et al.*, 2016). The moral dimension of biosafety and biosecurity of synthetic biology introduced above touches questions such as: How safe is safe enough; what risk is acceptable and according to which criteria; is it legitimate to weigh up expected benefits with unclear risks; are there knock-out arguments which morally forbid cost–benefit comparisons? Answers to questions of this type are usually contested and controversial. This type of normative uncertainty reflects diverging assessments of safety and risk (Sec. 4.3.1). Another normative uncertainty emerges out of the production of new living things (or technically strongly modified ones) by synthetic biology. It raises the question of the moral status of the artificial organisms and touches upon issues around the "dignity" of life (Sec. 4.3.2). Furthermore, metaphysical questions have also entered the game. In synthetic biology, humans move strongly from being modifiers of what is present to creators of something new in the dimension of life. This motivates questions of possible human hubris, or whether humans could be "playing God" by creating new forms of life (Sec. 4.3.3). In all of these fields, possible limitations to the development of synthetic biology, or recommendations of where it should go, must have external, e.g., ethical, origin:

> As such, synthetic biology does not comprise in itself restrictions with respect to scope and extent of its intervention; and it does not from within its own logic recognize an inherent value of living beings, their interests and behaviour (Boldt, 2014, 247).

4.3.1 Risks of Self-Organizing Products of Synthetic Biology

The openness concerning the future possible applications of synthetic biology render it senseless to apply established risk management. This would need to consider specific applications in specific areas of use, which is neither in the interest of this volume nor possible, given the early stage of development of most work in synthetic biology. Instead, we can build on the overview provided in Sec. 4.2.2 as a point of departure for reflecting on the types of risk that can be linked with normative uncertainty. The major character-istic in this regard is that synthetic biology concerns interventions in or constructions of *living systems*. These systems are in contrast to, e.g., abiotic synthetic nanoparticles (Grunwald, 2008a; Weckert and Moor, 2007) and are characterized by *self-organization*: they can reproduce and multiply, they can modify their properties (with-in limits) as a consequence of evolutionary pressure, and they can exchange substances as well as gene sequences with their environ-ment. Subsequently, a risk scenario concerning these synthetic or-ganisms takes on a completely different character than, for example, those about synthetic chemicals or synthetic nanoparticles.

The idea of self-organization has already served as an inspiration in the visions of a molecular assembler (Drexler, 1986), at the beginning of the history of nanotechnology. This assembler was supposed to be self-replicating and produce, as part of its propagation, further machines, which in turn could build further machines themselves. This idea, which Drexler considered to be the solution to practically all of humankind's problems, was reinterpreted by Bill Joy as a horror scenario in which the future would no longer need humans because civilization could continue without us (Joy, 2000). Some authors regarded it as at least conceivable that the creation of synthetic living and therefore self-organizing objects *can* pose a danger to the entirety of humankind in the sense described by Hans Jonas (1979). The fears presented in the following refer to the shared property of self-organization and are independent of whether we catalog the objects of fear as technology, biofacts (Karafyllis, 2008), or living things. The main question in this section will be whether and to what extent it is possible to infer results from the analysis about the role of the precautionary principle and related approaches for synthetic nanoparticles to the field of synthetic biology, and where the possible limits lie.

The famous work of Hans Jonas (1979) started with the observation that particular technical developments might have an apocalyptic potential threatening the future existence of humankind. According to his normative point of departure, that the existence of humankind must not be endangered, Jonas formulated a new, and in his option *categorical* imperative: "act in a way that the consequences of your actions are compatible to the permanence of real human life on Earth" (Jonas, 1979, 86). To make his principle more operable Jonas postulated a heuristics of fear in order to get an impression of possible negative developments and an obligation to use the worst scenario as orientation for action. If the worst scenario showed an apocalyptic potential then the respective action should not be taken. For synthetic biology, this approach would easily provide an argumentation to apply at least the highest safety standards, or even to ban synthetic biology because evil developments endangering the entirety of humankind are "thinkable" due to self-organization getting out of control, as a worst case scenario.

However, it might be possible to construct a worst case scenario showing an apocalyptic dimension for almost all technologies. Furthermore, banning the technology under consideration could also cause situations with unlimited risk. A situation of *aporia* would be the consequence, with an irresolvable internal contradiction demonstrating that Jonas' imperative of responsibility does not provide applicable and operable ethical guidance. The reason is that Jonas did not formulate any requirements concerning evidence for the worst case scenarios crucial for his argumentation. For Jonas, the mere thinkability of an apocalyptic worst case is sufficient to reject the whole line of technology. This positioning, however, opens up space for arbitrary speculation about worst cases, which lead to the situation of *aporia* outlined above. Hence, Jonas' approach might be appropriate for raising awareness with regard to situations involving possible dramatic risk. It is, however, inadequate as ethical guidance for action and decision-making (Grunwald, 2008b).

A wide international agreement on the Precautionary Principle, as a neighboring but different approach, was approved during the Earth Summit, the United Nations Conference on Environment and Development (UNCED), in Rio de Janeiro 1992, which became part of the Agenda 21 and orientated the formulation of the Sustainable Development Goals approved by the United Nations in 2015:

> In order to protect the environment, the precautionary approach should be widely applied by States according to their capabilities. Where there are threats of serious or irreversible damage, lack of full scientific certainty shall not be used as a reason for postponing cost-effective measures to prevent environmental degradation (Principle 15 of the Rio Declaration).

This principle was incorporated into the Treaty on the European Union in 1992 and is part of its environmental legislation (Article 174 of the Treaty; cp. Sec. 3.3.2). It establishes a rationale for political action without already having valid knowledge about adverse effects and substantially lowers the threshold for action by governments (von Schomberg, 2005). Waiting for certain and comprehensive knowledge would imply that political action always had to be postponed by the argument that scientific knowledge was still incomplete. In this way, preventive and protective action could simply come far too late. The precautionary principle opens up a way to act in cases of "valid suspicion" between the mere thinkability of adverse effects (Jonas), and the postulate of needing full and complete knowledge before acting at all (Persson, 2017) contra the laws of fear (Sunstein, 2005).

It is, however, a difficult task to clarify the meaning of "valid suspicion" or "reasonable concern" without either running into the possible high risk of a "wait-and-see" strategy (Gannon, 2003), or overstressing precautionary argumentation with the consequence of no longer being able to act at all. Thinking about the application of the precautionary principle (cp. Sec. 3.3.2) generally starts with a *scientific examination*. There is a need to have an assessment of the state of the knowledge available in science and of the types and extents of uncertainties involved. In assessing the uncertainties involved, it has to be clarified whether there is "reasonable concern" in this situation (Munthe, 2018). The question of normative guidance is transformed into a procedural challenge for assessing the status of the knowledge available (cp. also Giese and von Gleich, 2014; Winter, 2014). Monitoring of the changing status of knowledge and repeated assessments form the basis for a permanent learning process (Munthe, 2018), allowing the integration of new bio- and gene technologies step by step in a responsible manner (Ammann *et al.*, 2007). Accordingly, this process is an essential part of the risk governance (cp., e.g., recently von Gleich, 2020, for the field of gene drives at the occasion of SPAGE technologies).

Dupuy and Grinbaum (Dupuy and Grinbaum, 2004; Dupuy, 2007) have undertaken a trenchant interpretation, criticizing the precautionary principle as being too weak, therefore not allowing for responsible risk governance. They do not refer concretely to synthetic biology but to self-organizing nanotechnology. According to the authors, the growing significance of self-organization in technological approaches will inevitably lead to a catastrophe. They maintain that its complexity means that this development can be neither predicted nor controlled. Their argument consists in drawing attention to the necessity of the *deed* before it is possible for us to think about controlling its consequences:

> The unpredictable behavior of nanoscale objects means that engineers will not know how to make nanomachines until they actually start building them (*The Economist*, cited after Dupuy and Grinbaum, 2004, 288).

Regarding options for control, the authors continue: "The engineers of the future will be the ones who know they are successful when they are surprised by their own creations" (p. 289), and in consequence, "it is neither by error nor by terror that Man will be dispossessed of his own creations but by design" (p. 291). They conclude from this diagnosis that the classic techniques for dealing with uncertainty and unclear risks (Grunwald, 2008b) are insufficient: "We believe that none of these tools [utility analysis, scenario technique, precautionary principle, A.G.] is appropriate for tackling the situation that we are facing now" (*The Economist*, cited after Dupuy and Grinbaum, 2004, 293). More research will not necessarily increase certainty based on our knowledge but rather create more uncertainty:

> In cases where uncertainty is such that it entails that uncertainty itself is uncertain, it is impossible to know whether or not the conditions for application of the precautionary principle have been met. If we apply the principle to itself it will invalidate itself before our eyes (Dupuy and Grinbaum, 2004, 294).

The issue for the authors is instead to replace the open nature of the future, represented as tree diagrams on a linear time scale, by a *projected time* in which the connection between today's talk of the future and the bringing forth of a future present is established. The means of a self-fulfilling prophecy is to be used in this manner to determine the future. "We will call *prophecy* the determination of the future in projected time, by reference to the logic of self-fulfilling prophecy" (Dupuy and Grinbaum, 2004, 303). Even exceeding the gloom of Hans Jonas' "heuristics of fear" (1979), Dupuy and Grinbaum assert a duty to anticipate the catastrophe in order to prevent its occurrence:

> A catastrophe must necessarily be inscribed in the future with some vanishing, but non-zero weight, this being the condition for this catastrophe *not* to occur. The future, on this part, is held as real. This means that a human agent is told to live with an inscribed catastrophe. Only so will he avoid the occurrence of this catastrophe (Dupuy and Grinbaum, 2004, 310).

However, this interpretation fails for two reasons. First, it only leads to the same situation of *aporia* as the approach of Hans Jonas (see above), with the consequence that it is not helpful as ethical guidance for action. Second, it is built on a negative judgment of the precautionary principle, expressing a misunderstanding. The sentence that this diagnosis is built on is not tenable:

> In truth, one observes that applications of the "precautionary principle" generally boil down to little more than a glorified version of "cost–benefit" analysis (Dupuy and Grinbaum, 2004, 293).

The authors do not recognize that the precautionary principle is a *political* principle; neither a specification for a calculation following a preceding quantification, nor a legal specification simply to be applied (see the detailed presentation in von Schomberg, 2005; cp. also Ammann, 2007). The argument cited above, that application of the precautionary principle to itself makes it obsolete, is wrong. The opposite is the case: the precautionary principle was constructed and established precisely for this reason.[1] Beyond the precautionary principle, there is no room for a rational debate about possibly problematic futures influenced by synthetic biology under conditions of the highest or deepest uncertainty (Hansson, 2006). The limits to the rationally possible and negotiable are *already sounded out* in the precautionary principle itself. This means that the precautionary principle is the appropriate framework for addressing the risks of synthetic biology.

---

[1]This would undoubtedly cause specific problems in operationalization, which would increase the difficulties of management. Yet, practical issues are not an argument against the precautionary principle but point to a challenge to be met.

As a result, we can state that speculative narratives about possible catastrophic futures enabled by synthetic biology research are not appropriate for drawing sound conclusions to gain ethical guidance. They only run into arbitrary conclusions because of the situations of *aporia* they create. The precautionary principle avoids this situation and allows debate and decision-making about the next steps to be undertaken. It avoids thinking from an envisaged but merely speculative endpoint and instead considers possible next steps and postulates taking these steps in a responsible manner. We can therefore directly build on the experience gained by working with genetically modified organisms (Synth-Ethics, 2011; Breckling and Schmidt, 2014). In view of the facts that, compared to traditional gene technology, synthetic biology leads to a further increase in the depth of human intervention into living systems and that the pace of innovation continues to increase, the obligation for precaution may become even stronger. Responsibility will form a major issue in the run-up to genuine regulation and risk management in today's processes of research, and beyond (Sec. 4.4).

4.3.2   The Moral Status of Created Organisms

The production of new living things (or technically strongly modified ones) by synthetic biology will raise the question of their moral status. With respect to its moral status – and various bioethical positions differ on this considerably – a difference in principle is made between the living and nonliving objects of ethical reflection, so the question will be whether synthetically produced living things are also accorded this moral status. Would synthetically manufactured living things also be accorded reverence for life in Albert Schweitzer's sense? This implies first the question of whether a living thing in itself, i.e., for no other reason than that it is alive, is accorded a special moral status, such as dignity or a value of its own, as opposed to a value relative to purposes set by humans. According to the bio-centered position of bioethics, even the justification of technical manipulation of simple forms of life could be considered problematic, or it could be asked if synthetic living things are supposed to have the same moral status as natural ones:

Given that creations of synthetic biology can qualify as living organisms designed to be useful, can they be regarded solely as having instrumental value and being devoid of any intrinsic value (Bhutkar, 2005)?

There is a severe problem: if life has a value of its own, why should it be relevant whether it came into being naturally or technically? Each new form of life would add value to the world. The goal of synthetic biology is to *create* life, not to destroy it. It is therefore hardly imaginable that synthetic biology could come into conflict with a presumed intrinsic value of life. Maybe the mere act of creating life is morally questionable, but that is a different question and does not hinge on the intrinsic value of the results of this creation process (Link, 2010).

Another worry could be that we could lose respect for life in general by getting used to considering organisms as living machines (e.g., Boldt, 2016). But first, that would only be an indirect slippery slope argument based on mere speculation. Second, the question has to be raised whether "respect for life" is an appropriate and ethically legitimate attitude at all. In ethics, "respect for life" is not an issue because life is an abstract notion (Sec. 2.1), perhaps even a buzzword (Funk *et al.*, 2019). Rather, "respect for (particular, depending on the approach) *living beings*" is thematized, because living beings are specific entities. Therefore, the question of whether life *as such* has an intrinsic value, and if and how synthetic biology could come into conflict with it, can be excluded from further consideration because of its unclear meaning.

For many people however, the conviction that life has an intrinsic value is deeply rooted in their intuitions, often related to an implicit or explicit pathogenic approach to bioethics. This would require us to protect life for its own sake as soon as life can be exposed to harm and pain. But from which harm could or should one protect a cell or a bacterium? Provided that a living being lacks any ability for subjective experience, it is impossible to inflict pain on it. In other words, living organisms, e.g., bacteria, simply fail to have a relevant basis for being protected against harm (Link, 2010).

Bacteria and viruses – which themselves or their components are the objects of work in synthetic biology – are not the proprietors of rights in the customary ethical categories that could be harmed. In this sense, technical manipulations of these living things would not be ethically relevant, as manipulations *of them* but only with regard to the possible consequences of these manipulations *for animals or humans* (e.g., in the form of risks; see above). The same would apply, *a fortiori*, also to synthetic or technically modified living things, if they could be produced despite the objections mentioned above. Inasmuch as synthetic biology is concerned with living things that are not accorded any rights, ethical issues therefore are most strongly raised with respect to the *consequences* for humans and the environment, and not with respect to the technical manipulations of such life itself according to some presupposed dignity.

More far-reaching ethical questions are posed by the technical modification or creation of higher organisms. In particular, regarding the issue of the capacity of animals to experience pain, the focus of ethical consideration is not just on the consequences of technical interventions *for humans* but also on the affected organisms themselves (pathocentric position).

In particular, it would be a completely different challenge if it was to become possible to create artificial animals. What, for example, could be said about the moral status of an artificial cat produced in a lab? This is a completely speculative question, but we can already reflect on the reasons why we assign a specific moral status and value to animals. Today, we are used to applying the biological category of a species, and we assign a cat the moral status of an animal to be covered by animal protection rules because the cat belongs to the species of cats – and we know this because it was borne by another cat. But an artificial cat will not be borne by another cat; it will be manufactured in a lab. In a case where this artificial cat has exactly the same properties and abilities as a biological cat, many people will say: "Okay, let's consider it as a cat in the familiar sense because we cannot distinguish between this artificial cat and a natural one – they have exactly the same abilities and behavior." Others might say: "This is an artefact designed by humans, like a machine, therefore let's characterize it as a mere technical object, beyond the protection of laws on animal rights and welfare":

> The notion of "fabrication of life" is itself an example of this ambivalence, because it not only raises fundamental doubts about whether

life could be the object of fabrication but also the ontological question of whether the entities of synthetic biology are living beings or arte-facts (Funk *et al.*, 2019, 177).

We easily can see the resulting normative uncertainty and can prepare for possibly emerging decisions by anticipatory reflection.

Even if one does not take a strong position on animal ethics issues, this is evidently a field in which normative uncertainties can manifest themselves with regard to how higher organisms are handled as a consequence of measures of synthetic biology, or at least their possibility in a far future. The ground has been well prepared, however, with regard to how these uncertainties will be handled in the framework of ethics as a result of the debates over how animals are treated in the agricultural industry and of the ethical problems related to animal experiments and animal enhancement (Chap. 5).

4.3.3 Human Hubris by "Playing God"?

In synthetic biology, humans move from being modifiers of what is present to being creators of something new, at least according to the visions of some biologists:

> In fact, if synthetic biology as an activity of creation differs from genetic engineering as a manipulative approach, the Baconian *homo faber* will turn into a creator (Boldt and Müller, 2008, 387).

This does not necessarily seem revolutionary. Humans have always been creating something new, which was not part of the world before, e.g., synthetic chemicals or synthetic materials and nanoparticles. However, in the case of creating life, the perception is different. In 2005, even a high-level expert group working on behalf of the European Commission said it was likely that creating new life forms would give rise to fears of human hubris and of synthetic biologists "playing God" (following Dabrock, 2009). While the accusation that humans are playing God is presumably as old as technology itself, it is sometimes leveraged as a populist rhetorical maneuver against new technologies, and it implies mistrust in science. Often, it is fed by fundamentalist religious or spiritual beliefs, or by quasi-religious ecological convictions. But a substantial portion of the public still exhibits some reluctance toward allowing scientists to meddle with life, either by incubating it in a laboratory or by developing it from scratch.

The question should be scrutinized seriously, especially since "playing God" is one of the favorite buzzwords in media coverage of synthetic biology.[2] For example, an article in the influential German news magazine *Der Spiegel* (following Synth-Ethics, 2011) was titled "Konkurrenz für Gott" (Competing with God) in 2008, referring to the ETC group: "For the first time, *God* has *competition*" (ETC Group, 2007). The introduction to the piece stated that the aim of a group of biologists is to reinvent life, raising fears concerning human hubris (Grolle, 2010). The goal of understanding and fundamentally recreating life would, according to this article, provoke fears of humankind taking over God's role and that a being like Frankenstein's monster could be created in the lab.

Some ethicists and theologians have objected in general to any technology that attempts to create life on the grounds that we may not "play God." Yet it is unclear what this concern means argumentatively:

> There is very little agreement about the precise nature or legitimacy of these concerns, let alone what, if anything, might be done to address them (Parens *et al.*, 2009).

Most of the publications "that mention this class of issues usually do not elaborate them. Most papers are restricted to raising questions in a rather unsystematic sense" (de Vriend, 2006). These issues are often briefly mentioned in reports or articles containing an overview of ethical aspects of synthetic biology but are rarely scrutinized in-depth. An important task in achieving a fruitful debate would therefore be to clarify the ethical question: What does "playing God" mean?

Even the position that (some) religious people might fear "that when the creation of life at human hands happens this will knock down a key theological tenet that only God can create life from non-life" (van den Belt, 2009), is controversial. Even if we accept a (particular) religious view, why should the capability of creating

---

[2]The thoughts presented in this section have been elaborated in the Synth-Ethics project of the European Union (Synth-Ethics, 2011). In particular, I will refer to analyses made by Link (2010).

life threaten to undermine God's dignity or authority, at least as the "prime mover"?

> It is hard to credit the view that God would give us the abilities to make new life forms and then argue that to do so crosses a line that God does not want crossed (van den Belt, 2009).

It is of interest that this view is supported by many Christian, e.g., Roman Catholic theologians, in particular pointing to Genesis as the first book of the Bible. There, the story of the creation of humans by God is told in the following way:

> So God created man in his own image, in the image of God he created him; male and female he created them.

The notion of "in his own image" is interesting in our field. If God is the creator *per se* and He created humans "in his own image," then have humans also been created *as creators*? According to this reading, there is nothing wrong with humans acting as creators of something new, including new forms of life. Accordingly, at least Christian belief does not necessarily lead to fundamental problems with synthetic biology. It rather would address issues of responsibility for the process and products of research and development, as well as on innovation built upon them.

An analysis of the literature shows that there are two important things to be done. First, it would be ill-advised to simply push aside the worries and fears associated with phrases like "playing God." This is so if the aim is to ignite a public debate and to ensure there is social agreement on synthetic biology as a field of research. Therefore, we might first consciously examine what rationale may lay behind these fears, even if at first glimpse it seems to be based on nothing more than irrational uneasiness. Second, an examination of political philosophy may be needed to determine how to deal with these fears in democratic societies, if it turns out that many of these concerns are not well argued but have factual power (cp. de Vriend, 2006). Clearly, we cannot just disregard these worries. But mere uneasiness does not seem to justify restricting or over-regulating scientific freedom. An approach must be developed that takes such concerns into account without legitimizing exaggerated postulates.

However, the interpretation that the creation of life is God's prerogative and privilege is to be rejected. Even in the view of many theologians, to think that synthetic biologists overstep a crucial boundary rests on a fundamental misunderstanding of the term *creation*. Creation, it is argued, should not be considered merely as causal origination. It is not about the first link of a causal chain but about the continual reason of the whole chain. Creation is, in this view, a hermeneutic category for interpreting the existing world and revealing a specific view of it; causal categories are out of place. We humans may thus participate in God's creation as having been created "in his own image" but merely as *cooperatores Dei*. We simply do not have the potential for a *creatio a novo* and therefore cannot break into this sacrosanct, divine domain at all: "Man can principally not act like God" (Dabrock, 2009). In conclusion, major voices of religion seem to provide no reason to accuse synthetic biologists of "playing God," since they simply cannot do so. The only possible sin would lie in the arrogance of their presumption that they were able to "play God."

A different and important aspect of "playing God" is addressed by referring to human hubris. Because synthetic biologists have a powerful instrument at their disposal, the worry is that they might "play" with nature in a careless manner. Moreover, since the possible impact of new technologies emerging from synthetic biology probably will increase dramatically, the risks might seem so high and incalculable that pretending to have them under control as good as amounts to "playing God." While this concern is certainly warranted, both the possibility that the techniques for creating life could fall into the wrong hands, as well as the harm that might result from an accidental release of engineered microbes from the "right hands" are already covered by the debate about biosafety and biosecurity (see above) and the precautionary principle. In this respect, the charge of "playing God" does not add anything new (Link, 2010).

We can conclude that the "playing God" argument is an expression of concern and fear but that it does not touch on any ethical issues themselves in the sphere of synthetic biology: "[...] metaphors such as 'Playing God' and 'creating life' are taken as an indicator of social discontent and moral irritation" (Funk *et al.*, 2019, 180). Ultimately, there is no accepted normative framework enabling us to distinguish where "playing God" begins and where it ends. We similarly do not have the means to distinguish hubris from non-hubris. In problem-oriented ethics, there is no pathway for operating with the hubris or "playing God" position that could offer orientation for action. These are concerns, but they are not arguments.

In summary, we can concede that it is quite understandable to feel unease with regard to humans creating life. And the objections or demurs put forward can be fairly interpreted as attempts to conceptualize this unease. But all the issues considered, relying on the distinction between the natural and the artificial, or claiming that life has an intrinsic value, are not arguments apt to put this discomfort on a firm footing. This, of course, does not imply that these qualms are irrational or a-rational. Even if some of our intuitions rest on a "secularized religiousness," we are well advised to explore what the rationale behind them might be (Link, 2010), for two reasons:

- The speakers that advance these positions should be taken seriously as moral actors having something in mind which could also give rise to some concerns after ethical scrutiny.
- Previously invisible arguments might be hidden behind this uneasiness. Uncovering those possible items might be a task of searching for hermeneutic understanding rather than one of ethics but could also give rise to thinking about the possible normative uncertainties involved.

Therefore, discussions around the concerns of human hubris, or humans "playing God," could be a valuable source for ethics. These debates should be taken as empirical materials to uncover possibly hidden concerns, possible hidden normative uncertainties and ethical issues to be dealt with – but they are not arguments *per se*.

*

Summarizing briefly the results of this section makes clear that neither issues of the moral dignity of any products of synthetic biology (Sec. 4.3.2), nor the fear of humans "playing God" (Sec. 4.3.3) can give valid guidance on how to proceed with synthetic biology. The precautionary principle remains as a signpost toward identifying a responsible way to take the immediate and perhaps the next steps in research, development, and innovation. This implies carefully organizing responsibility assignments and the respective processes of assignment and exertion.

4.4 Responsibility Configurations of Synthetic Biology

The task is now to become more specific about actors influencing the further course of synthetic biology, their responsibilities and accountabilities, and about objects of responsibility beyond consequentialism, according to the EEE model of responsibility (Sec. 3.3.1).

Here, we first have briefly to recall the very nature of synthetic biology (Pade *et al.*, 2014; cp. Sec. 4.1). Characterizing it as a *technoscience* (Nordmann, 2014a; Grunwald, 2014b) has consequences for the assignment of responsibility, because the traditional boundary between technology-oriented applied science and cognition-oriented basic research is disappearing. The distribution of responsibilities in the field of synthetic biology has to take into account the broad range between basic science and possible applications.

Not surprisingly, the well-known conference of Asilomar held in 1975 is repeatedly referred to as a model for the responsibility of research and science in synthetic biology (e.g., Boldt and Müller, 2008). That conference took place in a situation in which a global spirit of optimism regarding genetic engineering could be observed, while at the same time the first signs of public criticism and demands for state regulation could be heard. As its outcome, the genetic engineering community committed itself to actively taking responsibility and exercising caution. However, interpretations of the conference are controversial. On the one hand, it is praised as a positive example of science proactively assuming responsibility; on the other hand, there are concerns that it mainly served the purpose of pre-empting state regulation, so that genetic engineering research could be carried on with as little interference as possible (de Vriend, 2006).

At the second global conference on synthetic biology in 2006, there was an attempt to follow up the actions taken at Asilomar. A corresponding declaration on the self-obligation of synthetic biology was passed (Maurer *et al.*, 2006). This, however, only refers to the possible *military* use of synthetic biology and to possible misuse. It lays out a set of self-obligations for synthetic biology in order to prevent possibilities of this kind ("Synthetic Biology/ SB2Declaration." *OpenWetWare.* https://openwetware.org/wiki/ Synthetic_Biology/SB2Declaration; accessed 30 Jan. 2020):

> First, we support the organization of an open working group that will undertake the coordinated development of improved software tools that can be used to check DNA synthesis orders for DNA sequences encoding hazardous biological systems; we expect that such software tools will be made freely available.
>
> Second, we support the adoption of best-practice sequence checking technology, including customer and order validation, by all commercial DNA synthesis companies; we encourage individuals and organizations to avoid patronizing companies that do not systematically check their DNA synthesis orders.
>
> Third, we support ongoing and future discussions within international science and engineering research communities for the purpose of developing creative solutions and frameworks that directly address challenges arising from the ongoing advances in biological technology, in particular, challenges to biological security and biological justice.
>
> Fourth, we support ongoing and future discussions with all stakeholders for the purpose of developing and analyzing inclusive governance options, including self-governance, that can be considered by policymakers and others such that the development and application of biological technology remains overwhelmingly constructive.

This manifest was heavily criticized by thirty-five civil society organizations, including the ETC Group, Greenpeace, and the Third World Network. The criticism addresses the perceived overly narrow perspective of the scientific self-obligation to prevent possible misuse and military use of synthetic biology and expresses mistrust against the instrument of self-obligation. The organizations postulate the need to assess and reflect on possible consequences of synthetic biology from a much broader perspective and to include social actors beyond science, such as stakeholders and citizens:

> Moreover, the social, economic, ethical, environmental, and human rights concerns that arise from the field of synthetic biology go far beyond deterring bioterrorists and "evildoers." Issues of ownership (including intellectual property), direction and control of the science, technology, processes, and products must also be thoroughly

considered. Society – especially social movements and marginalized peoples – must be fully engaged in designing and directing dialogue on the governance of synthetic biology. Because of the potential power and scope of this field, discussions and decisions concerning these technologies must take place in an accessible way (including physically accessible) at local, national, and global levels. In the absence of effective regulation, it is understandable that scientists are seeking to establish best practices but the real solution is for them to join with society to demand broad public oversight and governmental action to ensure social wellbeing (ETC Group, 2007).

This controversy points to the critical issue of determining an adequate relationship between science's autonomy and society's claim on involvement in the governance of science (Siune *et al.*, 2009). While science's autonomy is usually regarded with high value in basic research, society's voice and involvement is frequently required in applied research. Thus, the responsibility situation in synthetic biology seems to be ambiguous; as a technoscience, it does not belong to only one type of research. Therefore, the controversy denotes normative uncertainty, not necessarily in judging synthetic biology but in the process of how assessments and judgments should be made, thus referring to the empirical dimension of distributing responsibility among groups in society. First, however, it is necessary to reflect on the *objects* of responsibility in synthetic biology according to the epistemic dimension, in order to unfold responsibility into the empirical dimension.

Identifying and determining sensible *objects* for responsibility assessments and debates strongly depends on the epistemic dimension in the EEE model (Sec. 3.3.1). Future stories of the development of synthetic biology, of its useful outcomes to society, and of the consequences of the real-world uses of those products, systems, and services are highly uncertain, as in most NEST fields, which render the consequentialist scheme (Fig. 3.1) widely inapplicable. Therefore, assignments of responsibility in synthetic biology obviously cannot rely on anticipations of innovation paths and products in a consequentialist inquiry. The only possibility is to look at *currently ongoing* research and debates from the hermeneutic turn perspective (Sec. 3.3.3). The specific responsibility constellation of synthetic biology then includes, on the one hand, issues of *current research* needing good scientific practice and established moral standards and, on the other, far-ranging but highly speculative visions and expectations debated in science, politics, and society. This distinction opens up two types of options for determining appropriate objects of responsibility (cp. the analogy to objects in technology assessment, Grunwald, 2019a): (1) the *design and conduct of current research* and (2) the *design and content of current debates* on synthetic biology. Both options offer the opportunity to talk constructively and substantially about responsibility objects and the empirical dimension in the EEE model. Both draw from the current situation, rather than from issues of a speculative future society in which synthetic biology could or would have major impacts:

(1) Considering the *current research* on synthetic biology does not demand any far-ranging prospective analysis. Instead, it can be confronted, for example, with the concerns of bio-safety and biosecurity (Sec. 4.3.1), raising questions about their realization in currently ongoing research activities, or those to be designed for the next stage of research. Similarly, the opportunities, limitations, but also risks of a "do-it-yourself" technology can be considered, because synthetic biology in the "do-it-yourself" mode is a practice today (Sauter *et al.*, 2015). Considering research policy and agenda-setting in synthetic biology, the next research subjects can also be debated, along with decision processes and criteria. For example, perspectives, processes, and experiences of responsible research and innovation (RRI) can add to agenda-setting and to designing new research ideas and projects.

(2) Likewise, without a glance into the future, we can still debate visions for the future and possibly also other futures for synthetic biology, since these are voiced *today* and disseminated as interventions into an ongoing social debate. This debate ranges from expectations of salvation from the looming global energy crisis, e.g., by artificial photosynthesis, to the fear of "playing God" (see above). The more speculative the futures debated, the more significant are the conceptual, heuristic, and hermeneutic questions involved. The hermeneutic view on these future narratives (Sec. 3.3.3) would both clarify current debates as well as prepare for upcoming

debates, in which perhaps concrete technology will have to be assessed and evaluated. The hermeneutic approach is not only about understanding; it also opens up the opportunity to talk about whether and to what extent, and in what sense, intervening into ongoing debates with future narratives is itself responsible, or perhaps problematic (Grunwald, 2017). In the assignment of responsibility in synthetic biology, considering communicative acts as objects of responsibility is therefore of major importance – which would affect many actors, such as biologists, journalists, artists, science writers, policymakers, managers, civic organisations, and even private citizens in the empirical dimension.

In both directions, responsibility has to be assigned in a given *present* situation, with respect to present expectations and rules of assignment. However, there will be changes over time because of scientific and technological advances and changes in the societal environment and in awareness. In this change, the epistemic structure and the quality of available knowledge about consequences will also change, with implications for the appropriate objects to reflect upon. Responsibility reflections and assignments made during the very early stages of a development in synthetic biology will provide orientation for shaping the relevant *processes* of scientific advance and technological development, e.g., for the agenda-setting of science and reflective activities, for safety regulations, and for public dialogue. As the knowledge of synthetic biology grows, and with it the design and development of products and services, it will then be possible to continuously concretize the initially abstract orientations, motivating changes of assignments of responsibility. In this sense, responsibility reflection and assignment in all EEE dimensions is an ongoing process accompanying scientific and technological advance.

The following lists elements, which comprise possible *carriers* for the ethical reflection and assignment of responsibility for synthetic biology: persons and institutions to which responsibility could or should be assigned. It gives an impression of what the ethics of responsibility should include in this field, taking into account both the directions distinguished above (focussing on current issues and applying the hermeneutic perspective). The consideration of particular actors, institutions, and organizations allows us to talk of models of responsibility being shared among different groups, according to their respective accountabilities and opportunities for exerting influence. These elements are (Grunwald, 2014b):

- The *goals and objectives, even visions* of current research in synthetic biology. These could be confronted with questions of whether they are responsible, or could be made responsible by modifications (Grunwald, 2017).
- The envisioned, projected, or even merely imagined *products* of synthetic biology, in terms of materials, technological systems, and services based on knowledge provided by synthetic biology. These might include highly welcome outcomes, such as new and better drugs, but also problematic and unwanted developments such as biological weapons.
- The possible future *knowledge* of synthetic biology, which could influence not only our engineering capabilities and emerging innovation but also our understanding of life and of ourselves (Sec. 4.5).
- The consequences for *actor constellations and power distribution*: How could developments emerging out of synthetic biology influence power constellations and influence, e.g., in the related economies?
- The *science system*: We might ask about its ability and willingness to develop and establish reflective accompanying procedures to monitor and assess ongoing research in synthetic biology with respect to social, political, ethical, cultural, and other dimensions: are the preconditions of taking responsibility fulfilled by current structures and institutions in science?
- *Research funding*: Funding policies clearly influence the advance of synthetic biology. Therefore, the direction and themes of research funding in synthetic biology are subject to possible responsibility debates. In particular, facing scarcity of resources, current *priority-setting* in the allocation of financial and personal resources to synthetic biology research might be considered more or less responsible regarding other, perhaps more urgent, fields of research.

- The *legal and political framework,* which will influence the further advance and direction of research in synthetic biology, e.g., by regulation for risk governance (e.g., Winter, 2014) or incentive programs. These elements influence the ability to harvest benefits of synthetic biology and to prevent risks and hazards and are, therefore, also elements of responsibility for its further development.
- *Current research*: This might be assessed with respect to familiar responsibility criteria according to established guidelines for good scientific practice and codes of ethical conduct but also to existing regulation, e.g., precautionary measures, safety of researchers, observance of animal protection rules in animal experimentation, etc.
- The provision of knowledge-based and normatively reflected *policy advice*, which could also be seen as a subject of responsibility in this field, in order to support the development and adaptation of appropriate political boundary conditions with respect to public funding and regulation (e.g., Sauter *et al.*, 2015).
- Increasing society's *awareness* with regard to advances of synthetic biology and supporting an *open dialogue* about the further direction of research, which should be a subject of responsibility assignment, possibly also including public participation in the agenda-setting of synthetic biology (ETC Group, 2007).

This long but probably still incomplete list shows a wide variety of different types of possible objects to be reflected in responsibility assignments. Partially, these are directly linked with specific actors; partially it is not clear to whom which aspect of responsibility should be assigned. The variety of subjects, in combination with the variety of actors possibly made responsible, opens up a broad field of debate about legitimate, adequate, effective, and efficient distributions of responsibility in our field of consideration. Obviously, developing a comprehensive responsibility theory of synthetic biology would go far beyond the scope of this chapter.

The list presented above clearly illustrates that the empirical dimension of responsibility assignments in synthetic biology is heterogeneous. It will involve different types of actors, reaching from the biologists themselves to regulators, funding agencies and policymakers, through to civic organisations and private citizens. Assignment of responsibility must be based on normative pictures of how society should work, how humans should act as moral persons, and how science should serve society, e.g., on ideas of a deliberative democracy or on ambitious concepts of the modern governance of science in society (Siune *et al.*, 2009; cp. IRGC, 2009 for synthetic biology). In particular, it becomes clear that there is no "one-fits-all" solution to responsibility assignments. Rather, these assignments will lead to complex arrangements of distributed responsibilities, also indicating a division of labor and functional differentiation in this respect. Assigning responsibility to particular groups and expecting them to solve all the challenges (e.g., regarding engineers as moral heroes, Alpern, 1993) will remain under-complex and cannot meet the complexity of the challenge.

4.5 Beyond Ethics: Visionary Narratives Around the Notion of "Life"

As we could see when introducing the issues of dignity of life (Sec. 4.3.2) and the fear of "playing God," it is the notion of life, which makes people sensitive. This sensitivity is not automatically correlated with assumptions about possible unwanted consequences of synthetic biology but rather refers to assumptions of an intrinsic value of life.

> Synthetic biology can give rise to euphoric utopian scenarios, as well as to frightening dystopian narratives. These scenarios and narratives are not based on a consequentialist analysis of synthetic biology as a value neutral means to given ends (Boldt, 2014, 235).

Intervening into matter, as is done by nanotechnology in previously unknown depth, and intervening into viruses and the building bricks of living organisms are perceived as categorically different activities. But this difference can be understood in another manner:

(1) Life involves the capability of self-organization. Viruses can replicate, modify, and self-organize while rocks cannot. This difference causes different perceptions in terms of risk (Sec. 4.3.1).

(2) Life carries metaphysical connotations deeply rooted in human history, in the various religions, and in culture. Life is frequently assigned an intrinsic value or even some dignity, which is not assigned to, e.g., crystals and rocks. This issue goes further than the risk issue.

In both directions, narratives underpin and unfold the underlying assignments of meaning to life as such. The relation of particular forms of life to humans can also be made an issue at the conceptual level. Techno-visionary narratives are present in the debate on synthetic biology at different levels (Synth-Ethics, 2011). They include visions provided and disseminated by scientists and science promoters and visions disseminated by mass media, including negative visions stretching as far as dystopian views. They include stories about great progress in solving the energy problem, or contributing to huge steps in medicine, but also severe concerns about the possible non-controllability of self-organising systems (Dupuy and Grinbaum, 2004), or the already mentioned narrative of humans "playing God." As stated above (Sec. 3.3.3), there is epistemologically no way to clarify today whether these narratives tell us something sensible about the future or not. Therefore, we can only take the narratives, including their origins, and the intentions and diagnoses behind them, their meanings, their dissemination, and their impacts *as* the empirical data, and ask about their role in contemporary debates, renouncing any attempt of anticipation (Nordmann, 2014b). Critical hermeneutics (Roberge, 2011) offers to investigate the "stories behind the story."

For example, take the debate on "playing God" outlined above. Independent of the position that there is no argument behind this debate (Sec. 4.3.3) it should be scrutinized seriously, especially since "playing God" is one of the favorite buzzwords in media coverage of synthetic biology. Respective stories of human hubris provoke fears of humankind taking over God's role. This narrative is a dystopian version of the Baconian vision of full control over nature (Chap. 9). The hermeneutic approach means understanding what such debates could tell us, e.g., by reconstruction of the arguments and their premises used in the corresponding debates or by a historical analysis of the roots of the narratives used. Therefore, we have to look at the narratives relevant to synthetic biology with their diverse messages (following Grunwald, 2016a).

Many visions of synthetic biology tell stories about the paradise-like nature of scientific and technological advances in this field. Synthetic biology is expected to provide many benefits and to solve many of the urgent problems of humanity, primarily in the fields of energy, health, new materials, and more sustainable development (Sec. 4.2). The basic idea behind these expectations is that solutions which have developed in nature, could be made directly useful to human exploitation by synthetic biology:

> Nature has made highly precise and functional nanostructures for billions of years: DNA, proteins, membranes, filaments and cellular components. These biological nanostructures typically consist of simple molecular building blocks of limited chemical diversity arranged into a vast number of complex three-dimensional architectures and dynamic interaction patterns. Nature has evolved the ultimate design principles for nanoscale assembly by supplying and transforming building blocks such as atoms and molecules into functional nanostructures and utilizing templating and self-assembly principles, thereby providing systems that can self-replicate, self-repair, self-generate and self-destroy (Wagner, 2005, 39).

In analyzing those natural systems solutions and adapting them to human needs, the traditional boundary between biotic and abiotic systems could be transgressed. In the narrative "nature as a model for technology" (Grunwald, 2016a) humans will develop technologies following the models provided by nature, thereby expecting a reconciliation between technology and nature. In this narrative, the term "nanobionics" is often used. Bionics, often denoted as "biomimetics," attempts, as is frequently expressed metaphorically, to employ scientific means to learn from nature in order to solve technical problems (von Gleich *et al.*, 2007). The major promise of bionics, in the eyes of its protagonists, is that the bionic approach will make it possible to achieve a technology that is more natural or better adapted to nature than is possible with traditional technology. Examples of desired properties that could be achieved include adaptation into natural cycles, low levels of risk, fault tolerance, and environmental compatibility.

In grounding such expectations, advocates refer to the problem-solving properties of natural living systems, such as optimization according to multiple criteria under variable boundary conditions in the course of evolution and the use of available or closed materials cycles (von Gleich *et al.*, 2007, 30ff.). According to these expectations, the targeted exploitation of physical principles, of the possibilities for chemical synthesis, and of the functional properties of biological nanostructures is supposed to enable synthetic biology to achieve new technical features in hitherto unachieved complexity, with nature ultimately serving as the model.

These ideas refer to traditional bionics and biomimetics, which aimed (and aims) at learning from nature (e.g., animals or plants) at a macroscopic level. Transferred to the micro- or even nano-level it takes on an even more utopian character. If humans become able to act, following nature as the model, at the level of the "building bricks" of life, then an even more "nature-friendly" or nature-compatible technology could be expected. Philosophically, this is reminiscent of the idea of the German philosopher Ernst Bloch, who proposed an "alliance technology" (*Allianztechnik*) in order to reconcile nature and technology. While in the traditional way of designing technology, nature is regarded as a kind of "enemy," which must brought under control, Bloch proposes to develop future technology in accordance with nature, in order to arrive at a status of peaceful co-existence between humans and the natural environment.

Thus, this narrative related to synthetic biology is not totally new but goes back to early philosophical concerns about the dichotomy between technology and nature. The postulate related to this narrative, however, suffers from the fallacy of naturalness, which takes naturalness as a guarantee against risk. Nature is regarded as positive simply because it is nature. However, it is easy to tell a narrative of synthetic biology from completely the opposite direction, despite being based on the same characteristics.

Based on traditional Baconian reasoning, synthetic biology can be regarded as the full-blown triumph of Bacon's "dominion over nature" utopia (Sec. 9.1). The idea of controlling more and more parts of nature continues the basic convictions of the European Enlightenment in the Baconian tradition. From that perspective, human advance includes achieving more and more independence from any restrictions imposed by nature or natural evolution, to enable humankind to shape its environment and living conditions according to human values, preferences, and interests to a maximum extent.

The cognitive process of synthetic biology attempts to gather knowledge about the structures and functions of natural systems, not from contemplation or via distanced observation of nature but from *technical intervention*. Living systems are not of interest *as such*, for example, in their respective ecological or aesthetical contexts, but are analyzed in the *relationship of their technical functioning*. Living systems are thus interpreted as *technical systems* by synthetic biology. This can easily be seen in the extension of classical machine language to the sphere of the living, which is increasingly being described in techno-morph terms (cp. Sec. 4.1). Thus, the main indicator of the relevance of this understanding of synthetic biology and its meaning is the use of language. Synthetic biology is linked epistemologically to a technical view of the world and to technical intervention. It carries these technical ideas into the natural world, modulates nature in a techno-morph manner, and gains specific knowledge from this perspective. Nature is seen as technology, both in its individual components and as a whole:

> This is where a natural scientific reductionist view of the world is linked to a mechanistic technical one, according to which nature is consequently also just an engineer [...]. Since we can allegedly make its construction principles into our own, we can only see machines wherever we look – in human cells just as in the products of nanotechnology (Nordmann, 2007b, 221).

Instead of eliciting a more natural technology *per se* as promised by the bionic understanding of synthetic biology (see above), this research signifies a far-reaching technicalization of what is regarded as natural. *Beyond* considering synthetic biology from the perspective of ethics or societal debate and assessment, it appears sensible to ask if and how such changes in the use of language and such re-interpretations aiming at a different understanding modify the relationship between *technology* and *life* or modify our view of this relationship (see Chap. 9).

Chapter 5

Animal Enhancement for Human Purposes

The contemporary origin of the notion of "enhancement" with respect to humans (Chap. 7) and, subsequently, to animals (this chapter) can be found in the idea of the so-called NBIC convergence. Nanotechnology, biotechnology and genetic engineering, information and communications technology, and cognitive science and brain research are expected to converge, according to the often-cited report of the National Science Foundation (Roco and Bainbridge, 2002). This convergence is supposed to open up radically new opportunities, the focus of which, in that report, is on "improving human performance" (Chap. 7). The notion of "animal enhancement" (Ferrari *et al.*, 2010) was seemingly transferred from the idea of human enhancement, to the field of animal enhancement for the benefit of humans and human use, aiming at improving the usefulness of animals for human purposes (Bateman *et al.*, 2015).

This chapter first describes some of the key directions of research that are currently in progress in this field (Sec. 5.1). It then inquires about the semantics of *animal* enhancement (Sec. 5.2), in part in preparation for considerations of *human* enhancement (Chap. 7). Against the backdrop of existing normative frameworks, emerging ethical challenges are identified (Sec. 5.3), followed by reflections on responsibility for animal enhancement (Sec. 5.4).

Living Technology: Philosophy and Ethics at the Crossroads Between Life and Technology
Armin Grunwald
Copyright © 2021 Jenny Stanford Publishing Pte. Ltd.
ISBN 978-981-4877-70-1 (Hardcover), 978-1-003-14711-4 (eBook)
www.jennystanford.com

Final considerations lead beyond the level of applied ethics to the hermeneutics of the changing relationship between humans, animals, and technology (Sec. 5.5).

5.1 Technologies for Intervening into Animals

The use of animals for human purposes, such as for food or as working animals, has a long tradition in human civilization and our relationship with nature. It takes different forms, such as experimental animals in science, working and farm animals in agriculture, in the military, as pets, in zoos, and for leisure and entertainment purposes or for sports. In order to help humans utilize animals as effectively as possible, techniques for keeping and breeding animals were developed early in human history and have been developed further over time. Genetic engineering and molecular biology have now enabled profound, often also contested, steps toward further-reaching interventions into the nature of farm animals (Eriksson *et al.*, 2018). It appears that nanobiotechnology, gene editing, and other advanced biotechnologies will enable humans to take additional steps to increase this influence over animals.[1] The term "animal enhancement" is used occasionally in this field (Bateman *et al.*, 2015). However, its meaning needs to be clarified (Sec. 5.2).

Traditional techniques of breeding and keeping animals gradually changed the relation between humans and nature. Since the 1970s, new technologies have allowed deeper interventions into animals (Shriver and McConnachie, 2018). They draw especially on existing knowledge and experience in genetics, molecular biology, and veterinary medicine, which involve mostly incremental processes rather than disruptive ones. More recently, the idea of NBIC converging technologies (Roco and Bainbridge, 2002) has fueled expectations of making much deeper interventions into life possible. The NBIC convergence assumes that nanotechnology,

---

[1]This chapter is based on a study prepared for the Swiss Federal Ethics Committee on the Non-Human Field (Ferrari *et al.*, 2010). I would like to thank Arianna Ferrari, Christopher Coenen, and Arnold Sauter for permitting me to present some of the joint results here. Special thanks are dedicated to Arianna for valuable hints for revising and modernizing the current chapter compared to its predecessor (cp. also Ferrari, 2018).

biological and genetic technology, information and communications technology, and cognitive science and brain research will converge and unify in order to become a new technoscience. Nanotechnology as the science of dealing technologically with matter at the level of the smallest particles, including atoms and molecules, provides key competences required for this convergence (Schmid *et al.*, 2006; Schmidt, 2008). Convergence tendencies can be recognized which will also be observed in the animal realm, such as the enrichment of research and technology in molecular biology by the use of methods from neurophysiology and the linking of information and communications technology with genetic procedures. In this development, synthetic biology (Chap. 4), nanobiotechnology, and gene editing enable humans to intervene faster and much more deeply into the nature of animals. These developments legitimate speaking of "animal enhancement" as a new step in the relations between humans and natural animals. In the following, several illustrative fields of research will be introduced, which demonstrate opportunities and potential for deepened interventions into animals in order to better enable meeting human interests.

The utility of animals dominates the perception of what constitutes the "enhancement"[2] of farm animals, e.g., by gene editing. Agricultural research focuses on maximizing animals' performance primarily for economic reasons. In this area, animal "enhancement" often dovetails with classical forms of "enhancing" farm animals in agriculture. The specific purpose of cloning farm animals, for example, is to reproduce animals that have already been "enhanced" by breeding in order to further improve and multiply their utility. This combination of classical approaches with new technological means also includes visions regarding converging technologies, such as the use of intelligent biosensors in farm animals that can permanently check the state of an animal's health in order to systemically integrate diagnosis and drug delivery (Roco and Bainbridge, 2002, 5f.). In this regard, nanoelectronics, nanobiotechnology, and bioinformatics are expected to play a central role, since extremely small instruments have to be produced that can activate and regulate themselves (Scott,

---

[2]The word "enhancement" will, as a matter of principle, be placed in quotation marks in the following. This is to indicate that it is important to note the substantial conceptual complexity of the word in contrast to its simplistic positive connotation (Sec. 5.2).

2005). The production and utilization of transgenic farm animals is also based on the desire for an improved or even perfect design of animals with regard to human interests. A recent example addressed the problem of cow's milk causing allergies in infants. Researchers have successfully edited a cow's genome so that the cow will produce milk free from a specific allergen responsible for that allergy (Wei *et al.*, 2018; Zhaolin *et al.*, 2018).

Increasing animals' competitiveness is the goal of the use of doping medication and of cloning sport animals for breeding. Cloned horses are used as suppliers of valuable reproductive material, such as sperm. The direct doping of animals, in contrast, is forbidden by the regulations of sporting associations but has been reported, especially with regard to racehorses. Conditions such as strong competition, insufficient regulation and control, and economic interests can lead to illegal "enhancement" by doping, just as in doping in human sports (Waddington and Smith, 2008). However, it is not well known how widespread such practices are.

New technology also makes "enhancements" possible in animals kept as pets, in response to new human desires and needs. Cosmetic surgery is used to modify individual animals to meet the respective aesthetic desires of humans. The goal might be to make the animals more successful in animal beauty pageants, or simply to make them more attractive to their owners. Even though it might be advantageous for a dog to be more attractive to humans if it then were treated better, this does not change the fact that the operation takes place according to human criteria and intentions only. The assumed advantage for the dog would simply be a side effect of increasing its utility to humans, rather than the primary goal. Fish also provide examples of the dedicated "enhancement" of pets. The fluorescent zebra fish Night Pearl from the Taiwanese company Taikong has been on the market in Asia since 2003. This is a genetically modified fish, in whose genome a gene for a fluorescent protein was inserted. The production and sale of these fish met with objections in Europe (Whitehouse, 2003). The trade in genetically modified fish was forbidden in Europe, Canada, and Australia during the late 2000s. The trade of transgenic salmon, however, was permitted in Canada in 2019. Further fluorescent fish products have in the meantime reached the market (Robischon, 2007).

The most frequent technological applications in the area of animal "enhancement" are animal models for dedicated experiments.[3] In biomedicine, animal models for human diseases or their symptoms are frequently produced in order to test therapies or medications prior to clinical trials. "Enhancement" of the experimental animals serves to improve the conclusiveness and applicability of the subsequent results for humans. Examples are xenotransplantation (Tisato and Cozzi, 2012, in which animals serve as the source of organs, tissues, and cells for humans, and the use of mouse models for investigating cause–effect chains and possible therapies for diseases such as Alzheimer's. To facilitate this, animals are enhanced in different ways. For example, pigs are modified so that their tissue elicits a milder rejection response in primates, or primates are treated pharmacologically so that they accept the porcine organs or tissues more readily (cp. Lu *et al.*, 2019, for recent developments).

Animals are also used in research activities for developing human "enhancements" (Chap. 7), e.g., for testing medication or implants. Animal experiments are also taking place in the research and development of novel brain–machine interfaces. For rodents and apes it has been demonstrated numerous times that implanted electrodes make new means possible for the animals to control machines. Following training, apes with implanted electrodes are able to control robot arms sufficiently that they can grasp objects and feed themselves (Velliste *et al.*, 2008). Apes have been successfully trained to move their own temporarily paralyzed arm in a targeted manner, using a brain–computer interface (Moritz *et al.*, 2008). Technology was also successfully developed to remotely control animals. Rats' brains, for example, have been stimulated in such a manner that it was possible to precisely control their movements (Talwar *et al.*, 2002). The remote control of external objects by means of brain activity is thus a reality in animal models (Hatsopoulos and Donoghue, 2009).

Nonetheless, it is only possible to speak of animal "enhancements" to a very limited degree. In some cases, individual animals were previously paralyzed or injured in order to simulate the analogous situation of humans, or they were bred and prepared as disease

---

[3]The term "animal models" does not refer to models of animals but to real (often living) animals that are considered to be models for certain functional relationships and that therefore can serve as objects for dedicated experiments.

models. Examples include rat models of traumatic injuries to the spinal cord (Truin *et al.*, 2009) and of paralysis following stroke (Pedrono *et al.*, 2010), Parkinson's disease (Fuentes *et al.*, 2009), and Huntington's disease (Kraft *et al.*, 2009).

Animal experiments are also taking place in work on extending the human life span, which is a significant topic in human enhancement (Chap. 7). Of particular interest are mice that have been the object of various types of interventions. While it is true that life spans of experimental animals have been extended, it is very difficult to interpret the meaning of the results for humans or to transfer them to humans (Ferrari *et al.*, 2010) since it is unclear what is being compared.

A well-known example, which in a certain sense has led to a deep change in animals' properties, is the mouse model known as *fearless mouse*. In this mouse model, the gene encoding stathmin was knocked out (Shumyatsky *et al.*, 2005). Stathmin inhibits the formation of microtubules, which are responsible for the delivery of information about learned and innate fear to the amygdala, a central area of the brain important for memory. Such mice have a reduced memory of fearful experiences, so they cannot recognize dangerous situations because they lack the congenital mechanisms. Transgenic mice have also been created, whose olfactory system has been modified so that they possess none of the congenital mechanisms for recognizing bad or dangerous scents. A mouse approaching a cat and even snuggling with it was reported (Kobayakawa *et al.*, 2007). It is obvious that knocking out fear does not necessarily constitute a form of positive improvement from the imagined perspective of animals, since under normal conditions the mechanisms of fear are decisive for survival. Researchers expect these models to produce new knowledge for treating human mental disorders such as panic attacks or post-traumatic stress disorder, as well as new insights into the mechanisms of fear.

Recently, there is increasing interest in brain organoid research. A brain organoid is an artificially grown, *in vitro* miniature organ claimed to function similarly to animal or even human brains. These can be created by culturing human pluripotent stem cells in a bioreactor. This development could open up further opportunities for deepening human intervention into the living world. Major ethical issues, however, concern procurement of human biomaterials and donor consent, the animal research required, a possible organoid consciousness, and the moral status of the results (Hyun *et al.*, 2020).

Summarizing, several research activities aim at "enhancing" animals. This is happening for a range of diverse purposes and is applied to different species. The main fields are using animals for experiments and tests for measures targeted to humans as well as considering the utility of animals in agriculture. "Enhancements" for sports and pets also exist, but not to the same extent. In general, the new converging technologies, such as nanobiotechnology and nanoinformatics, and new methods in the field of genetic manipulation such as gene editing, are used to deepen human intervention into the nature of animals.

5.2 The Semantics of Animal Enhancement

The term "enhancement" has fundamentally positive connotations in everyday language as the opposite of "diminishing." Often, it is used in the sense of an improvement *per se* with an intrinsically positive meaning. The word "enhancement" and its use are, however, semantically much richer and more complex than is often assumed. Therefore, it must be examined carefully in order to avoid confusion and misunderstanding and to prevent running into the rhetorical trap suggested by everyday language. Rather, we have to ask about the criteria according to which a simple change observed or aimed at is regarded positively as an *enhancement* or *improvement.*

"Enhancement" or "enhancing" represent activities or actions for influencing particular objects to travel in a desired direction. With respect to grammar, we can distinguish between subjects and objects: there are *actors* (the *subjects* of enhancement) who enhance *something* (the *object* of enhancement) according to *criteria* which describe the intended direction of change and which may be used to measure the success or failure of the intended enhancement.

Already this simple reconstruction allows us to apply the difference between goals and consequences well known from action theory. While the *goals* of an enhancement measure are envisaged *ex ante,* the actual consequences of the measure show themselves *ex post* only. A measure intended to be an enhancement can turn out *ex post* to be a dimunition, or it can have other unintended consequences. In an ethical scrutiny we, therefore, have to consider the *goals* of the enhancement and the *criteria* oriented toward these goals as well as the possible *unintended consequences* of the enhancement measures.

While this reconstruction fully covers the grammar of enhancing, its semantics needs a more differentiated picture. Considering the dynamics of enhancement, we have to talk not only about the direction of desired change but also about its *point of departure* and an appropriate measure of the *extent* of the change gained. The dynamics of "enhancement" necessarily includes three semantic dimensions:

(1) A *starting point* for the enhancement process. An enhancement is only plausible as an *enhancement* if the starting point of the change can clearly be stated and will allow observation of change.

(2) A *criterion* of enhancement. Normative criteria guiding the enhancement in the intended direction must be determined. A criterion consists of the declaration of a *parameter* (quantitative or qualitative) and the *direction* in which the parameter is to be altered in order to create the intended enhancement.

(3) A *measure* of enhancement. Measuring the *extent* of an enhancement as the difference between the starting point and the status reached so far is primarily significant in order to be able to assess the speed of enhancement. It will also be important in weighting processes, if the intended enhancement in one respect is offset by deterioration in another and balancing is necessary.

I will use the term "animal enhancement" to refer to the following practices:

(1) Practices that are supposed to enhance the performance of animals compared to a starting point of "established" performance. The understanding of "performance" usually will be oriented to human purposes, i.e., users' interests in making use of the animals.

(2) Practices that permit or facilitate the human use of animals in which animals are exposed to fewer or no heavy burdens than would be the case without the enhancement measures.

(3) Practices that include the creation of *new* properties in animals, such as resistance to diseases, that in a strict sense go beyond improving already existing properties or performances and thus constitute a change.

This definition covers the usual understanding of "enhancing" animals for human purposes but also includes the advocatory perspective. Therefore, it does not pre-determine results of any ethical assessment.

Many conflicts about possible enhancements, regardless of whether they concern humans or animals, flare up (1) due to the very criteria for enhancement, (2) due to different perspectives on the given enhancement, and (3) with regard to possible unintended side effects.

(1) Something desired or accepted by certain actors or groups as an "enhancement" could be considered problematic by others, which might lead to conflict and rejection. Different actors might prefer different criteria for "enhancement" due to different moral convictions. The criteria for designating a change as an "enhancement" are normative. They apply and attribute certain values, which may not be shared by others. As an example, statements such as, "It is sensible to enhance farm animals with respect to their efficiency in agriculture," may not meet with universal approval or with the values of people engaged in animal welfare.

(2) In connection with the criteria employed, there is also a *social* dimension of assessing and judging possible enhancements. The question can (and often should) be raised of *for whom* an "enhancement" is truly an enhancement. An example is the debate about genetically modified plants in the production of food. "Enhancements" such as resistance to herbicides for increasing production efficiency would primarily benefit agricultural producers and the food industry. Possible undesired effects on health would however land on the consumers of GMO food. In cases of animal "enhancement," this configuration becomes even more complicated: the question of the consequences of that "enhancement" *for the animals*, and not only for the human actors who are the

subjects of "enhancement" and those who benefit, always has to be answered (Sec. 5.3.1).

(3) Furthermore, a change that is an "enhancement" relative to a particular criterion can be a dimunition relative to another. As is well known from technical development, the "enhancement" of a car from the point of view of sportiness can often only be achieved at the expense of burdening the environment, and an "enhancement" in terms of decreased costs can endanger safety requirements. Conflicting goals are an everyday issue in engineering but also in political decision-making. "Enhancements" in general are not improvements *per se*. While some properties will be improved, others might deteriorate. This observation makes complex assessments of the relative value of "enhancement" *versus* deterioration and weighting of them necessary before a decision can be made.[4]

A further semantic consideration addresses the relation between the terms "enhancement," "perfection," and "optimization," which are often confused. *Enhancing* describes, as introduced above, a change compared to a starting point in an intended direction. It is open with regard to the extent of change without a definite endpoint. *Optimizing* and *perfecting*, however, are oriented at an envisaged *final* or *target status*. An "optimization" comes to an end as soon as the optimum is reached.

In this chapter, we will not look at *optimizing* or *perfecting* animals but at "enhancing" them. While these words often are used synonymously, they show a deep difference. "Enhancement" includes considering the direction and criteria of change without, however, aiming at achieving a perfect or optimal final status. This implies the absence of any *inherent* endpoint to the "enhancement": the result of each particular enhancement can serve as the starting point for new enhancements, possibly based on different criteria. The process of "enhancement" will never be completed. It rather opens up an infinite chain of possible "enhancements" in the future.

The application of the "enhancement" concept to animals involves several noteworthy challenges, which can be addressed

---

[4]This is also familiar from breeding. For example, the highest production cows are as a rule more susceptible to disease and less robust against environmental factors.

by adopting the semantic introduction of "enhancement" given above: "*someone enhances something according to a criterion*," and by using the references to the starting point, the criterion, and the extent of the "enhancement" mentioned above. While the objects of enhancements are obviously the animals under consideration, the subjects currently involved in "animal enhancement" can be separated into two groups:

- those who have intentions to enhance animals and determine the criteria of "enhancement." This will be, for example, the owners of the animals or the scientists using animals for experiments;
- those who practically conduct the "enhancements." These will be, for example, veterinary physicians, or researchers, who react in part to a demand from science and in part to a nonscientific demand for enhanced animals.

An anticipatory outlook toward responsibility issues (Sec. 5.4) demonstrates that it is not only the "makers" of "enhancement" that must be considered but also the actors determining the criteria for "enhancement." The *objects* of "enhancement" are animals or, more specifically, certain features of animals' performance.

Most interesting with respect to the cognitive interest of this volume is the question as to the criterion, or the multiple criteria, for "enhancement," the motivations and goals related to them, and the strategies for justifying them, as well as their limitations in a world that increasingly considers issues of animal welfare. This opens up the pathway for ethical reasoning.

5.3 Normative Frameworks and Ethical Challenges

The main tension in the field of "animal enhancement" consists of the conflict between desires to "enhance" animals to increase their utility for human purposes and taking care of animal welfare (Sec. 5.3.1). This tension affects most, perhaps all, of the normative uncertainties in "animal enhancement" (Sec. 5.3.2).

5.3.1 Between Human Purposes and Animal Welfare

"Animal enhancement" is not some future dream but both a part of current practice and an object of research (Sec. 5.1). Handling animals is a classic topic of ethics and is subject to underlying normative conditions and moral standards, which show concrete manifestations in animal welfare legislation. This legislation is different among various countries and depends on the respective moral status given to animals, which in itself depends on tradition, culture, and religion. While, for example, the concept of the "dignity of the creature" is anchored in the constitution in Switzerland and that of the intrinsic value of animals occurs in animal protection legislation in the Netherlands, there are no equivalent concepts in German law. In French law, animals are even considered to be things. The main differences concern the moral value assigned to animals – by humans. These differences do not necessarily lead to different animal protection rules among various countries. The normative framework of pathocentric ethics, which aims to avoid pain and suffering of animals, is widespread. However, in detail, and for analytical reasons, there is a pattern behind this: the weaker the rights of animals are anchored in legislation and culture, the more the character of intervening measures is tied to the desires and interests of animal owners.

Since about thirty years ago, a movement has become apparent that contains two different and contradictory features (Ryder, 2000). On the one hand, animal welfare is often attributed a high level of social significance, as expressed in public debate, in many initiatives and in increasingly tougher legislation, e.g., in EU legislation about cosmetic testing on animals. On the other, however, the pressure to use animals has risen, above all in the fields of agriculture and experimental animals. Developments in the life sciences, especially in genetic engineering and gene editing, and technological innovation in the course of the NBIC converging technologies have led to a greater depth of human intervention in animals, which leads to more critical questions as to the limits of such intervention. Because of the fundamental relationship to the context of usage by humans, the grand ethical issues of "animal enhancement" are similar to the existing ones, such as for which purposes animals may be instrumentalized, under what conditions animal experiments are permitted, and how we should deal with animals' stress and suffering.

The motivation for "animal enhancement" and its more specific goals are the points of departure for any action. Without intention to "enhance," there can be no "enhancement" at all. Therefore, we must look to human interests regarding animals to find answers to the question of the origin of the motivations for "enhancement." The examples presented in Sec. 5.1 provide us with a variety of motivations rooted in goals and purposes determined by humans for "enhancing" the utility of animals. However, we also have to consider the perspective of animals. The ideal way to legitimize the desire for an "enhancement," or at least to ease the legitimization, would be informed consent by the animals. However, in complete opposition to the field of human "enhancement" (Chap. 7) it is impossible to obtain informed consent from animals. "Animal enhancement" thus remains the product of human decisions. These can be dedicated to two different normative directions, usually leading to tensions:

- *Utilitarian perspective*: From this perspective, current or anticipated applications of animal enhancement happen such that animals are supposed to be utilized or better utilized for human purposes. Consequently, animal enhancement has to be justified in relation to human interests.

- *Advocatory perspective*: From this perspective, human arguments might be based on an imagined perspective of animals. Animal enhancement could be dedicated to goals of realizing these imagined and presumed animal interests, such as reducing pain, or leading a life appropriate to the respective species.

"Animal enhancement" can thus be understood, on the one hand, as a direct enhancement of the utility of animals to humans but, on the other, as enhancement in the presumed animal's interest, within certain practices of human usage. Both goals are set by humans, but against different normative frameworks devoting different attention to animal rights and welfare. This duality permeates the ethical issues of "animal enhancement." Usually it involves deep ambivalences and conflicts, because measures enhancing utility will not automatically serve the imagined interests of the animal. Mostly, the opposite might be the case.

Interventions to improve disease resistance or a genetically engineered change to optimize certain cognitive abilities of animals could appear to be an exception, or a different case. Although the creation of disease resistance should *prima facie* be in the imagined interests of animals, the situation is more complex. As a rule, the motivation to create disease resistance is not to save the animals from illnesses but to improve the opportunity for human use, as an example from agriculture illustrates. Among dairy cows, mastitis (inflammation of the udder) is, after fertility disturbances, the most frequent reason for premature slaughtering. A typical Holstein Friesian cow, the most widespread breed, today lives for less than five years, which is an unnaturally short life span. This also poses a problem for their owners since these cows do not reach the age of maximum performance (Reuter, 2007). "Enhancement" with respect to milk production has in this case increased susceptibility to illness. If the attempt were made to counteract this by means of a technical enhancement, i.e., to find a resistance to mastitis, this would – if successful – spare animals illness from the disease and premature slaughter. Yet the motivation for this "enhancement" would continue to be humans' interest in using farm animals more efficiently. "Animal enhancement" would ultimately be employed to compensate for negative developments (increasing mastitis) that are caused by the intense use of animals by humans in the first place. The talk about "enhancement" of animals' abilities in the assumed interest of the animals therefore appears to be ambivalent; it is about resolving problems which occur only because of prior intense utilization of these animals.

A key group for taking care of animals and also "animal enhancement" is veterinarians. They operate, in the first place, within the normative framework of their respective national legislation, which can differ considerably. In the second place, codes of conduct or ethical guidelines of veterinary associations contribute to the profession's moral orientation. In contrast to human medicine, there is no ethos of healing animals as an end *in itself*. Instead, healing usually supports achieving human interests in animals in the different fields of their use. Because of the different moral and legal statuses of humans and animals, human and veterinary medicine are subject to different underlying normative frameworks.

In spite of the comparably weak moral position of animals, such codes of conduct or guidelines can provide some orientation. As an example, we consider veterinary interventions into pets in response to the needs of their human owners, e.g., by cosmetic surgery to modify animals to meet particular aesthetic desires. The assumption that it might be advantageous for a cat to be more attractive and beautiful because it will then probably be treated better, should not obscure the fact that surgery takes place according to human criteria only. Such interventions have been criticized as unnecessary medical actions according to a veterinarian's ethos (Neumann, 2008). In order to underpin this judgment, the ethical principles of the American Veterinary Medical Association (AVMA, 2020) tells us in Article I:

> §I: A veterinarian shall be dedicated to providing competent veterinary medical care, with compassion and respect for animal welfare and public health.

> §I,c: The choice of treatments or animal care shall not be influenced by considerations other than the welfare of the patient, the needs of the client, and the safety of the public.

> §I,e: Performance of surgical or other procedures in any species for the purpose of concealing genetic defects in animals to be shown, raced, bred, or sold is unethical. However, should the health or welfare of the individual patient require correction of such genetic defects, it is recommended that the patient be rendered incapable of reproduction.

This prescription is a clear obstacle to some types of "animal enhancement." However, the basic issue remains open: "Animal enhancement" always occurs in the tension between animal rights and human interests (Shriver and McConnachie, 2018). In any case, intervening into animals must be considered ethically, with regard to (1) the *goal* of intervention as the intended endpoint of a measure, concerning e.g., the desired performance properties. An ethical judgment also must take into consideration (2) the entire *process*, including the research and the interventions that must take place on route to creating an "enhanced" animal. It is not sufficient to observe the properties of an "enhanced animal – its well-being, integrity, and similar features – but the route to the goal must also be taken into account. Furthermore, (3) possible *side effects* have to be considered in advance of an intervention, as far as possible.

5.3.2   Ethical Spotlights on Animal Enhancement

In this section, some recent discussions in the areas of animal experiments and the extensive change with regard to animals' freedom from pain will be reported in order to demonstrate the nature of ethics involved in "animal enhancement" and to prepare for the deeper philosophical questions involved (Sec. 5.5).

Animal Experiments

Animal experiments in research have long been a matter of controversy. Under what conditions may animals be exposed to pain and suffering in order to gain knowledge (Shriver, 2015)? An important aspect concerns the different estimates concerning the necessity of performing research on living models and the possibility of avoiding *in vivo* experiments in order to reach the goals of toxicological and biomedical research. Fundamental arguments in this debate concern the following assertions (Nuffield Council on Bioethics, 2005):

- The differences between the species mean that transferring the results gained from animals to humans is problematic.
- It is challenging to solve the problem of selecting the appropriate species for a particular experiment.
- A reductionist understanding of illnesses is a precondition for animal experiments because at the most only a few physiologically similar phenomena can be represented in animal models.

Advocates of the use of animal models employ primarily two arguments to support their position. The first is that studies on living organisms are an essential part of biological and biomedical research because of the similarities between species, which as a matter of principle cannot be completely replaced by *in vitro* studies. The second is that it follows from the necessity of *in vivo* studies that animal experiments are not only acceptable but even ethically desirable because they make it possible to avoid experiments on humans (DFG, 2004). Responding to the problem of the difference between species, these advocates point out that the level of empirical knowledge has grown substantially and is currently becoming more precise due to biostatistical procedures.

Genetic engineering, gene editing, molecular biology, and nanobiotechnology make possible new types of research and interventions, such as xenotransplantation or gene pharming, which also lead to a need for further animal experiments. In this way their introduction repeatedly stimulates the debate about the suitability of animal models and their limits. For example, the assumption that rapid and massive progress could be made from the creation of knockout lines of mice was an object of increasing controversy in the scientific community (Qiu, 2006). The criticism raised was not a matter of ethics but was motivated, in terms of the theory of science, by the question of the anticipated gain in knowledge compared to the substantial funding required for this purpose. In this way, scientific progress always raises new demands to adapt the existing normative framework for animal experiments to other cases and to make it more adequate and specific. The fundamental issues, however, remain the same.

A new dimension in the evaluation of the procedures of modern research from the perspective of animal ethics is a consequence, however, of the interaction between human "enhancement" (Chap. 7) and "animal enhancement." Animal experiments make up much of the work on animal enhancement (Ferrari *et al.*, 2010), which is frequently in preparation for human "enhancement." While animal experiments are ethically widely accepted if their results can be used therapeutically for humans, the situation is completely different if the goal changes from human therapy to human "enhancement," because human "enhancement" is by definition not therapy (Sec. 7.2). If human "enhancement" is classified as a luxury, then animal experiments for this purpose would be similarly problematic to those conducted for cosmetics, for example. There is an urgent need for ethical and juridical clarification, in particular if – as, e.g., in the Swiss regulations on animal experiments – there are legally regulated conditions for permitting animal experiments. Exclusion criteria for animal experiments such as those for designing new cosmetics, which have already been defined in some countries for genetically engineered changes to animals, could also be applied to research and development for enhancement technologies. This challenge to the existing normative framework must urgently be made the subject of ethical debate at the societal level and, if appropriate, to subsequent legal clarification.

Elimination of Animals' Capacity for Suffering

Aiming at the reduction of the pain and suffering of animals is among the core convictions of animal ethics in the pathocentric paradigm. Following the AlleyDog.com Psychology Glossary:

> Pathocentrism is a philosophical position regarding the difference or similarities in the reaction to pain in humans and animals. As a philosophical stance, there are several varied opinions and positions taken on the question as to whether humans and animals suffer pain differently and whether that is due to differences in mental capacity and sentience (awareness of things). Pathocentrism is a philosophical and ethical discussion that is a major consideration in the study and implementation of bioethics and the use of living beings in experimentation ("Pathocentrism." Alleydog.com's online glossary. https://www.alleydog.com/glossary/definition-cit.php?term=Pathocentrism; accessed 21 Sep. 2020).

The issue of reducing or entirely eliminating animals' capacity for suffering has been discussed in the context of the evaluation of far-reaching interventions in animals.[5] If animals no longer suffered, then pathocentric ethics would no longer apply to them (Shriver, 2015). Considering the description of "animal enhancement" given above as an improvement relative to the expectations for human usage, this could in fact be interpreted as "enhancement" if subsequent human use of animals was simpler because one would no longer have to take an animal's well-being into consideration, thus making human use of animals more efficient in an economic sense. This may initially sound cynical. From an imagined perspective of animals one could say, if animals are going to be used for human purposes anyway, then it is better for them not to suffer. But before continuing this difficult discussion, I would first like to mention a concrete example for illustration.

Animal microencephalic lumps (AML) have been an object of discussion for some time. As a result of genetic engineering, these

---

[5]To be quite clear, these considerations are thought experiments and do not describe elements of ongoing research.

hypothetical animals would no longer be in a position to feel any interests or needs in any form and would thus be incapable of suffering. AML chickens, for example, would lack extremities and be featherless, and they could not see or hear or move (Comstock, 2000). From AML pigs and AML cattle it would be possible to regularly remove pieces of meat for consumption since they would grow back, assuming that appropriate nutrients are provided. Laying hens could be modified by genetic engineering so that they would not suffer any longer in their tiny cages. This manner of thinking can also be applied to experimental animals: if experiments must inevitably be carried out on animal models, this could lessen the suffering.

AMLs would be reduced to the status of beings incapable of experiencing sensation; an extreme reduction. This could be regarded as a maximal instrumentalization of animals, in which their "animal character" would largely or completely disappear, leaving behind a type of "utility object." Such an intervention would basically be so all-encompassing that almost all of the qualities of the animals would be changed. Questions would be raised such as whether AMLs are still animals and whether they are therefore still subject to animal welfare regulations, or whether they were not rather biofacts (Karafyllis, 2008), in which the *created* portion so eclipses the natural that the latter no longer plays any role. However, if we understand animal welfare holistically as the complex entirety of an animal's physical and mental states, then the loss of the possibility to achieve positive mental status, such as following a massive reduction in cognitive abilities, must be seen as a decrease in their integrity (Holtug, 1996).

The creation of such animals does not appear to be absolutely improbable or purely fantastic. Developments in this direction do not seem completely absurd if we think of the genetically engineered changes in farm animals to make them better adapted to the conditions under which they are kept. Making them disease resistant, removing the visual ability of chickens, or the creation of so-called zombie mice in the context of research on the neuronal correlates of consciousness (Koch, 2004), are examples which demonstrate that some researchers and research funders, which are highly relevant to the further course of research development, are thinking in that direction. At the same time, these considerations are still speculative.

The extreme scenarios of AML farm animals and transgenic mice made incapable of sensation could evoke counter-reactions. On the one hand, they could be rejected intuitively in that such animals might be seen as monsters; the repulsive products of technology. Intuitively, such a manipulation of animals appears to be illegitimate. On the other hand, such rejection might subside precisely because these beings would no longer be viewed as animals in the sense of fellow creatures but rather as living beings such as plants, or even as simple collections of cells and tissue that are incapable of sensation (Schmidt, 2008). This could even be regarded as a shift of the ontological status of those animals.

The question of the maximum reduction in animal attributes and in particular the elimination of their capacity for pain poses a special challenge to animal ethics because the capacity to suffer constitutes a direct source of moral relevance within pathocentric ethics (see above) and is also an important criterion for other theories of animal ethics (Ferrari, 2008). Although many people think that intensive animal farming should be forbidden as soon as it is associated with pain and suffering for animals, the situation is: if we continue to live in a world in which meat is consumed, it is better for every measure to be undertaken to avoid this suffering or to minimize overall suffering (Shriver, 2009). To replace current farm animals by other kinds of animals to be bred that have a lower or no capacity for pain would lead to better consequences than the status quo. From this techno-centered perspective, the elimination of the capacity for pain could be regarded as justified not only by human interests but also by an imagined interest of animals (Attfield, 1995). A notable counterargument is, however, that the creation of AML animals would be morally reprehensible not because harm would be done to the AML animals themselves but because it is impossible to imagine such animals being created without a larger number of *other* animals experiencing suffering and pain (Comstock, 2000; Ferrari, 2008).

This discussion points to the significance of protecting the welfare and the integrity or even "dignity" of animals as well as to the limitations of a fixation on animal pain and suffering. Ultimately, the issue is the moral status of animals and the consequences of "enhancement." A central question is whether ethical evaluation should be based on the capacity for pain in the pathocentric paradigm, or on the totality of the species-specific attributes of the animal covered by concepts such as integrity and possible inherent value. The answer will determine the standards for determining, which and how many interventions in animals are acceptable. It will also influence which criteria the evaluation should employ, which in turn will determine the appraisal of far-reaching "animal enhancements."

Transgressing boundaries between humans and animals

Transgressing the boundary between species, such as between humans and animals, is made possible or facilitated by converging technologies. It is even the declared goal of xenotransplantation, with reference to the removal of animal organs or tissue for transplantation into humans (Lu *et al.*, 2019). Thinking further about these developments, and toward the creation of chimeras, out of the narrow therapeutic context of xenotransplantation leads to questions concerning the possible ethical problems and limits (Jorqui-Azofra, 2020).

Many of the questions about research on chimeras are similar to those in the discussion of genetically engineered modifications of animals, such as the question of whether a species, or an individual member of a species, has an intrinsic value, and therefore whether transgressing the boundary between species would be morally problematic or unnatural, and what this would mean (Robert and Baylis, 2003). The possible creation of human–animal composite creatures and in particular the possibility of novel types of instrumentalization of human living beings leads to the question of the moral and judicial status of possible chimeras (Chakrabarty, 2003). Fundamental questions that would be raised by the emergence of human–animal composite creatures are: How much animal material is necessary for a human to no longer be a human? Can a threshold value be determined? What criteria should guide the distinction between whether such creatures would be protected by the International Declaration of Human Rights or would fall under animal protection laws? In Denmark, for example, animal experiments are forbidden where they would lead to the creation of chimeras that strongly resemble humans (Danish Council of Ethics, 2008).

The rejection of human–animal composite creatures is based on two main arguments. One is that membership of the human species is the criterion for attributing human dignity, which is grounded either in religion or in anthropocentric metaphysics and which is protected, e.g., by the German constitution. Consequently, the mingling of the animal genome with that of humans would signify a violation of human dignity (Karpowicz *et al.*, 2005). The other argument is that humans have a special position because of their specific cognitive capacities, in particular their ability to think abstractly, to form complex thoughts, to use language, to reflect on possible futures, and to take responsibility (see, e.g., Höffe *et al.*, 2002). These arguments are not unchallenged. Recourse to membership in the human species proves to be circular, or a mere defense of anthropocentrism without its religious grounding, which is abandoned as a common ground in secular society (Ferrari, 2008). Furthermore, cognitive capacities can be seen on a continuum between humans and animals as new ethological studies show (e.g., Warneken and Tomasello, 2006). Therefore, it is problematic to ban mixed creatures with respect to factual properties of humans.[6]

In confronting the arguments against the creation of human–animal composite creatures, the Danish Council of Ethics (2008) suggested an argument that is relevant to reflection on the moral status of "enhanced" animals in general. A violation of the "dignity" of creatures would happen if, for example, an animal whose cognitive capacities have been "enhanced" is not kept in a manner appropriate to its ("enhanced") status. Turning this argument around means that "enhancement" can also result in *extended* moral obligations in dealing with the "enhanced" animals. However, again the meaning of the "dignity" of animals and its applicability remain contested.

---

[6]Assigning moral status to an entity by referring to its factual properties will frequently run into problems. In the chapter on synthetic biology, we discussed the thought experiment whether an artificial cat with identical properties and capabilities to natural cats would fall under animal protection laws, or would be regarded as a technical object like a washing machine.

5.4   Responsibility Considerations

"Animal enhancement" always occurs in the tension between animal rights and human interests. Therefore, conducting "animal enhancement" in a responsible manner is a complex challenge. On the one hand, there are social debates, analyses from philosophy and animal ethics, and efforts by animal welfare organizations in the form of civil groups. On the other, there are the concrete interests of the users of animals in science, agriculture, sports, the military, and entertainment. This duality leads to a complicated constellation of arguments and actors that takes on many forms. Analysis and debate are needed to clarify what should be understood by responsibility in the field of "animal enhancement" at the occasion of each measure. Possibilities of attributing and distributing concrete accountabilities and responsibilities have to be discussed and determined. In this section, only a few more general responsibility considerations can be elaborated, because going deeper into the issue would require more specific cases.

First of all, we have to concede that responsibility assignments for "animal enhancement" are confronted with a deep asymmetry between humans as subjects and animals as objects of the "enhancement." Humans define the purpose and direction of the interventions, while animals are affected by the measures taken. The Golden Rule of ethics, or any other postulates fundamentally grounded in convictions of symmetry and equity among humans do not apply here. The asymmetric relation between humans and animals shows itself also in the field of animal rights, because these have to be determined by humans. Even if we talk philosophically about the possible *inherent* rights of animals, human culture and legislation implementing these rights differ over times, cultures, and nations. The animals themselves do not and cannot have a voice in making laws or rules of how to respect animals in cases of their "enhancement."

The "interest" of the animal is a construct based on human interpretation. In spite of the fact that there are often good or even very good arguments for these interpretations, animals themselves are not able to clearly express their interests, preferences, and values. Any instrument comparable to the "informed consent" approach of medical ethics does not work because humans and animals do not share a discourse community (*Diskursgemeinschaft*, Habermas, 1988). Therefore, in the face of this asymmetry, responsibility assignments in this field necessarily have to be grounded in the ethics of care.

Following the concept of ethics used throughout this volume (Sec. 3.2), intervening into animals must be considered ethically with regard to (1) the *goal* of "enhancement," (2) the entire *process*, including the research and the measures to be taken for creating an "enhanced" animal, and (3) possible *side effects* such as failed "enhancement" efforts. While the goals of the "enhancement" refer to properties of the envisaged "enhanced" animals and might question their quality of life compared to non-enhanced animals, the process of how to get there evokes questions of research and animal ethics related to the measures taken and the instruments used. This scheme allows us to roughly structure the field of actors to whom responsibility can be assigned.

Determining the goals for any "enhancement," and even deciding that "animal enhancement" is wanted at all, is made by humans according to specific purposes, ends, expectations, intentions, and so forth. These actors are usually owners of animals, e.g., in agriculture or in sports, or are owners or shareholders of companies in the food and meat industry. Their intention is, formally speaking, to increase the utility of animals for their own purposes, whether this is better performance in sports, or higher efficiency in the production of meat or other products to be delivered by animals. The normative frameworks behind this are the ideals of fair competition in sports and business and utility maximization in food production. These frameworks are widely accepted in modern society, while simultaneously producing moral conflict regarding animal welfare issues. This type of conflict as well as the related responsibility issues are well known, e.g., from the field of environmental and climate problems related to consumption. While industry and its lobbyists usually point to the presumed needs of consumers, e.g., in the field of contested SUV vehicles and their role in non-sustainable mobility, environmental activists and organizations point to the economic interests of shareholders. From its structure, the responsibility debate looks similar in the field of "animal enhancement." It is the old question of the driving forces, the drivers, and the driven in human affairs, which often leads to thinning, sometimes even to seemingly disappearing responsibilities or to endless debates with one actor pointing to the responsibility of others, and vice versa.

Nevertheless, two directions of responsibility reflections result out of this conflict. First, the reasons brought forward for "animal enhancement," its rationale or even necessity, have to be scrutinized for their justifiability, which is a task in the ethical dimension of responsibility. Economic reasons are legitimate but reach their limits as soon as basic rights or protection rules could be violated. The tension between economic interests and animal welfare must be negotiated based on an ethical weighing up of the diverging arguments. The results depend on cultural and moral priorities. The stronger the emphasis on animal rights in a society, the more the promoters of "enhancement" will have to observe animal welfare in their actions.

Second, we have to ask where the claimed pressure to further increase economic efficiency in agriculture and the food industry comes from. On one side, competition in the global food and meat market is a strong driving force (Ferrari, 2017). Reducing the power of this force would probably decrease stress on animals, with or without "enhancement." In this respect, responsibility lies at the level of international bodies such as WHO but also with nation states negotiating free trade zones or agreeing on bi- or multilateral rules. However, the desires of consumers concerning meat must not be overlooked. Annual meat consumption has been increasing for decades, usually related to economic growth in more and more countries. In addition, large parts of the population, e.g., in Germany, take the price of meat as a main criterion of consumption and look for the cheapest meat available. Without any doubt, this type of low-cost meat consumption contributes to the stress on animals. It could be reduced by lower (or no) meat consumption, which would also contribute to solving sustainability challenges such as climate change and overuse of grasslands and waters.

All these configurations also play a role for responsibility considerations in "animal enhancement." However, they are not specific to our field of interest because they apply to making use of animals for meat consumption more generally. Looking at the *processes* of "animal enhancement" allows for clearer determination of responsibilities. Obviously, those implementing "animal enhancement" have to be regarded as responsible actors: researchers and technicians, animal care personnel, veterinarians, lab assistants have the responsibility to take care of animals during the processes of "enhancement." This responsibility applies *today*, and for the respective *next steps*, without challenges due to the epistemic problems of consequentialism (Sec. 3.3.2). An important issue during the processes of "enhancement" is also to anticipate and avoid, where possible, negative side effects causing suffering and pain to animals. In particular, in the field of animal experiments, when, e.g., innovative drugs or invasive techniques are explored, unexpected effects can occur more easily. Examples are known from human plastic surgery: there is no guarantee that interventions aiming at "enhancement" of beauty will succeed. Similarly, measures of "animal enhancement" could fail in some circumstances. The animals affected must then be treated in a responsible manner – what this will mean, depends on the case and can range from stopping the intervention early enough to a eugenic solution in the worst case.

While all these responsibility considerations are not surprising, I would like to raise an additional question motivating a completely different type of responsibility in the field of using and "enhancing" animals for meat production. If we accept consumers' desires for large quantities of cheap meat, we could ask whether increasing stress on animals and their "enhancement" is the only way to meet these desires. Is exploiting animals for increased meat consumption really the only way, or could we look for other sources of proteins: cribs, microalgae, or *in vitro* food? Part of further responsibility is looking for alternatives to reach the same goals. My thesis is that responsibility includes the development of alternatives to "enhancing" animals for mass production of cheap meat.

An example of diverging attitudes and emotions as well as normative uncertainty in the talk about "animal enhancement" is current research in the area of *in vitro* meat. At the end of the 1990s, NASA conducted experiments to create artificial tissue to provide food for space expeditions. In 1997, the animal protection organization PETA began to co-finance a project between research centers in the Netherlands and North America with the goal of growing animal-like tissue with the taste, texture, and muscle mass of fish or chicken meat. An international research community now exists whose motivation is not solely animal protection – namely for *in vitro* meat to replace traditional practices that are frequently fraught with suffering for animals – but also environmental protection, since the animal-based food industry is responsible for substantial CO_2 emissions, the deforestation of tropical forests, and water pollution (FAO, 2006). Furthermore, economic advantages are expected because the costs of keeping animals and damage to the environment would be eliminated. Finally, health protection is a factor, because "meat" produced in a laboratory would be subject to more rigorous testing and free of antibiotics and other animal medications. The production of *in vitro* meat is therefore tied to positive visions that could lead to sustained consumption without causing increased stress on animals (Böhm *et al.*, 2018). Because of the potential of this approach, my conclusion is that funding and further research to exploring *in vitro* meat is justified as one of possible pathways to improve sustainability as well as animal welfare (Woll, 2019).

5.5 Beyond Ethics: Changing Human–Animal Relationships

In dealing with all of these issues, it has become evident in the preceding sections that ethics is only *part of the game*. As in other cases of new and emerging developments in science and technology (e.g., Chap. 4), the issues around "animal enhancement" are in some regards not concrete ethical ones at all but rather affect human self-understanding and worldviews. Questions such as the development of the relationship of humans to animals and to the tendencies toward technicalization that are at work in this context need first and above all public dialogue and expert discourse. Until now, debates on the relationships between humans and animals have only taken place sporadically, for example, with regard to the conditions under which farm animals are kept, or the need for laboratory animals. The topic of the technicalization of animals has hardly reached the public or politics.

The fact that humans compare and distinguish themselves to and from animals is a decisive element in their self-reassurance, the determination of their own role and of the ontological position of animals. Animals, animal images, animal figures, and animal imitations have played a central role in magical and religious traditions since early human history. In the relatively recent monotheistic religions, many of the regulations for dealing with animals and the metaphorical use of animals in religious scriptures are of great significance. The machine model has gained influence in the self-determination and self-description of humans only in modern times. The Cartesian interpretation of the human as an animated machine agrees with an interpretation and depreciation of the animal as an inanimate machine. In today's visions of artificial intelligence and autonomous technology (Chap. 8), the relationships are changing as advanced machines are becoming closer to humans (van Est, 2014), if we define ourselves by our cognitive capacities, which can be imitated by technical systems to an ever stronger degree. The mutual comparisons between humans, machines, and animals are an essential element of debating humans' views on the order of the world and on their own position in that order.

Technology is increasingly penetrating our cohabitation with other species and modifying the material basis of the relationship in various ways. Culturally, we can note a stronger acknowledgment of animals as cohabitants of the earth, an increase in animal welfare, and a growth in civic involvement for a vegetarian or vegan diet and against animal experiments during the last two decades. On the other hand, there is a continuing spiral of increasing exploitation of animals, and "animal enhancement" is in this tradition. The feeling of being related to animals is increasing alongside the instrumentalization of animals for human ends. Currently, no clear developmental trend can be recognized in these contrasting observations. Monitoring, hermeneutic questions as to the significance of changes in the relationship between humans and animals, and thorough reflection on these analyses have to be addressed.

The view of current and anticipated future research in the context of "animal enhancement," as presented above, shows that new technologies have substantial consequences for the opportunities for the use of animals. The possibility of transgressing species boundaries, the extreme modification of animals' attributes, the increase in and intensification of animal experiments, and the introduction of artificial items into living beings give us occasion to fundamentally reconsider our relationship to animals. Converging technologies and gene editing allow deeper interventions into animals that motivate speaking of "animal enhancement." Their primary interest is not *understanding* nature but intentional *intervention* into nature and the creation of instruments for the purpose of intervention. From the perspective of converging technologies, nature, including living beings, is increasingly regarded as a complex configuration of fundamental components and processes at the level of atoms and molecules. The knowledge gained through techno-scientific investigation of living organisms from this perspective can be used for redesign, intervention, and "enhancement" (e.g., Ferrari *et al.*, 2010).

Fully understanding *animal enhancement* needs us to observe two conceptual steps. First, living beings have to be described in techno-morph terms similar to machines (similar to Sec. 4.5). Second, animals, as they are, are considered to be *inefficient* machines that are to be "enhanced" and *can* be "enhanced." However, their characterization as inefficient can only be a judgment made by humans based on their own expectations. "Animal enhancement" thus increases the technical part of animals created by humans and reduces the natural one.

Ideas of being able to achieve anything in the field of life also arise inasmuch as humans are viewed as the creators and controllers of these developments:

> The aim of this metaphysical program is to turn man into a demiurge or, scarcely more modestly, the "engineer of evolutionary processes." [...] This puts him in the position of being the divine maker of the world (Dupuy, 2005, 13).

These obviously immodest ideas regarding feasibility are fed by atomic reductionism, according to which all activity in the world can be traced back to causal processes at the level of atoms. This reductionism and the famous slogan of nanotechnology "Shaping the World Atom by Atom" (NNI, 1999) form a leitmotif (Nordmann, 2007b, 220ff.). If nanotechnology were to offer the possibility of controlling these processes technically, then humans would, so to speak, have control of the roots of all causal chains and could thus practically control everything. This interpretation envisages humans' ultimate triumph, namely that armed with nanotechnology humans could begin to manipulate the world atom by atom according to their own conceptions, which is ultimately the consummation of Bacon's thoughts (Schmidt, 2008). In deterministic thinking, control of the atomic dimension also means control over the spheres of life and of social phenomena – and also over the shapes of animals (Ferrari, 2015).

It is arguable whether this will change the human–animal relationship, or whether this has already happened to a large extent, indicating a colonization of nature (Ferrari, 2015). It is possible that radical shifts are also looming in the relationship between animal and machine – the two factors central for human self-understanding – in the form of animal–machine hybrids (Ferrari *et al.*, 2010). Methods and visions in this area could in the future raise the question of whether strongly modified animals – in particular, insects – could or should be regarded as animals any longer. Without recourse to these expectations of the future – or even to the visions of an artificial insect brain that seem fantastic – the degree of coalescence between animal and machine can already be interpreted as the sign of a new quality. And robots that can be controlled by the use of animal elements such as rat neurons also point in the direction of the de facto disintegration of the boundary between machine and animal. Thorough observation of these possible changes is indicated because of the strong cultural significance of the human–animal relationship. This, admittedly, is not a genuine task of ethics but rather one of philosophy of life, anthropology, hermeneutics, and the philosophy of technology and nature (Chap. 9).

Chapter 6

Shaping the Code of Human Life: Genome Editing

Genetics and gene technologies have been developed since the 1960s, following the discovery of the double helix DNA structure as the "code of life" by Watson and Crick in 1953. By importing computer language, it is also denoted as the "source code of life." The development of these technologies was and still is accompanied by often heavy debates on ethical issues (an early voice: Anderson, 1972). These comprise; moral permission for the manipulation of life at its core, the possible risks involved, the responsibility of interventions into the human germline, the emergence of a liberal eugenics (Habermas, 2005), the question of humans taking over control of evolution, and so forth. The conference of Asilomar in 1975 (Berg *et al.*, 1981) is often mentioned as the first major event visible in the public domain, where scientists debated responsibilities in this field.

Currently, the situation shows itself as differentiated. The field of genetically modified plants and GM food seems to be frozen, with high levels of acceptance in some parts of the world and strict rejection and corresponding regulation in others. Genetic modification of animals, usually to enhance them for realizing human purposes (Chap. 5), is

Living Technology: Philosophy and Ethics at the Crossroads Between Life and Technology
Armin Grunwald
Copyright © 2021 Jenny Stanford Publishing Pte. Ltd.
ISBN 978-981-4877-70-1 (Hardcover), 978-1-003-14711-4 (eBook)
www.jennystanford.com

widely used and partially accepted. Intense ethical debates address gene technologies applied to humans, usually dedicated to medical purposes and in the area of reproduction. The most recent wave of ethical reasoning in this field followed the availability of the CRISPR-Cas9 genome editing technology, which makes interventions into genomes both faster and cheaper than earlier approaches, with better accuracy than was achievable in early genetics (e.g., Baumann, 2016; Araujo, 2017).

In this chapter, I will focus on genome editing applied to humans, and in particular on germline interventions. This type of intervention shows some specific ethical facets not covered by other chapters of this book. Taking the story of the Chinese researcher He Jiankui as an example (Sec. 6.1), I will first introduce some basics of genome editing technology and its envisaged fields of application (Sec. 6.2), then consider related normative uncertainties and ethical issues (Sec. 6.3), and the resulting issues of responsibility (Sec. 6.4). Finally, I will raise the question whether genome editing corresponds to editing *humans*.

6.1   Editing Humans? The Case of the Chinese Twins

In recent years, huge progress has been made in genetic engineering through the so-called genome editing technologies, in particular based on the CRISPR-Cas9 system. This breakthrough (Sec. 6.2) has been followed by a considerable research boom, with applications in plant and animal breeding as well as in human medicine. With respect to the latter, an expansion of therapeutic options based on genetic intervention is expected. In addition, and much more challenging in ethical respects, modifications of the germline as permanent interventions in the human genome reached the stage of first scientific experiments. As soon as the first reports on corresponding experiments with human embryos conducted by Chinese scientists were published in 2015, an intense debate among scientists and ethicists emerged. At first, this spread into political and public debate slowly, but this changed dramatically in late 2018 (Caplan, 2019).

In November 2018, the Chinese medical researcher He Jiankui reported to a scientific conference and to the public that he had created genetically edited human babies for the first time in history. The aim of his experiment was to insert HIV resistance into their genome in order to prevent HIV infection, either by their parents or in later life. Directly after the presentation, in mostly Chinese media, his report was regarded as a major scientific advancement and a historical breakthrough for disease prevention. Considering his experiments more closely, however, the perception changed quickly, and He received widespread and severe criticism (e.g., Cyranoski, 2018; Lovell-Badge, 2019; Greely, 2019; Begley, 2019; Lander *et al.*, 2019). His research activities were completely stopped by the Chinese authorities. He was fired and brought to court, received a penalty of three years imprisonment and a fine of three million Yuan. This story fueled a worldwide debate on the ethics of genome editing, based on earlier debates about genetic engineering of humans in general and interventions into the germline in particular (e.g., Habermas, 2005).

He's clinical experiment was organized as follows (Li *et al.*, 2019). His team recruited eight couples, each with an HIV-positive father and HIV-negative mother. During *in vitro* fertilization, the sperm was purified of HIV. Using CRISPR-Cas9 technology, the team mutated a gene in the genome of the embryos, which should cause/effectuate resistance to HIV infection. Criticism was raised because it was not clear whether the experiment had received proper ethical review from an independent board and whether the parents had really given fully informed consent (Krimsky, 2019; Li *et al.*, 2019). As the outcome of the experiment, twin girls with the pseudonyms Lulu and Nana were born in October 2018 and appeared to be healthy. However, it was and still is unclear if there could be adverse long-term effects in their future, caused by genome editing. It is unknown who should pay for medical treatment and care in case of serious health problems emerging during the lifetime of the children. Because of confidentiality, knowledge is not publicly available about the monitoring of the health status of the children, neither about processes nor about results. The experiment was conducted without any preparatory or accompanying discussion in the scientific and ethics community. In addition, He was criticized for having not really been interested in HIV prevention for the children but more in the experiment as such and in the subsequent worldwide publicity.

Some strong voices and statements of bioethicists on the irresponsibility of this event have been expressed (see also Begley, 2019; Li *et al.*, 2019). According to *Wikipedia* ("He Jiankui." *Wikipedia.* https://en.wikipedia.org/wiki/He_Jiankui; accessed 21 Sep. 2020), He's research and clinical experiment was regarded as: "a practice with the least degree of ethical justifiability and acceptability" (Ren-zong Qiu); "[...] a highly irresponsible, unethical, and dangerous use of genome editing technology" (Kathy Niakan); and "[...] monstrous" (Julian Savulescu). After the public report given by He on his clinical experiments, the Scientific Ethics Committee of the Academic Divisions of the Chinese Academy of Sciences declared their opposition to any clinical use of genome editing for germline cells, noting that

> [...] the theory is not reliable, the technology is deficient, the risks are uncontrollable, and ethics and regulations prohibit the action (quoted after Li *et al.*, 2019, 33).

A few voices, however, defended at least some aspects of the experiment (e.g., Gyngell *et al.*, 2019). In particular, its goal of using genome editing for creating HIV resistance and thus to prevent HIV infection was regarded as justifiable, since HIV is a huge threat to global public health. An even more positive judgment was made by thinkers close to transhumanism, such as Arthur Caplan. In his opinion, engineering human genes is inevitable and medical researchers are legitimately interested in using genome editing to prevent diseases, as He did. In April 2019 however, the Chinese Academy of Science (CAS) noted:

> [We] believe there is no sound scientific reason to perform this type of gene editing on the human germline, and that the behavior of He [Jiankui] and his team represents a gross violation of both the Chinese regulations and the consensus reached by the international science community. We strongly condemn their actions as extremely irresponsible, both scientifically and ethically ("He Jiankui." *Wikipedia.* https://en.wikipedia.org/wiki/He_Jiankui; accessed 19 May 2020).

In the meantime, Chinese regulations with respect to germline interventions have been tightened considerably (Cyranoski, 2019), as China demonstrated its will to stick to international consensus and ethical conduct.

The experiment with the gene-edited twins is the most recent story of scientific advance related to a singular situation fueling far-ranging and heavy normative uncertainty, debate, and conflict. When the first *in vitro* fertilized baby, Louise Brown, was born in 1978, a fundamental ethical debate emerged on *in vitro* fertilization in general. Today, this technique is well established in reproductive medicine, in spite of remaining problems. As soon as the first mammal, Dolly the sheep, was cloned in 1996, an intense debate began, not only on the cloning of animals but also of humans, which are also mammals. In the meantime, cloning of animals has become widespread and accepted in some areas, e.g., in cloning transgenic animals for experiments in pharma research (Chap. 5), but *reproductive* cloning of humans has been rejected at the international level and is forbidden by law in many countries (this is not the case for *therapeutic* cloning). The events related to the names Louise, Dolly, and the Chinese twins Lulu and Nana, were personifications and manifestations of preceding and sometimes long-lasting lines of research. As singular manifestations, however, they served as starting points for emerging ethical and public debates, in particular on the manipulation of human life. These ethical debates presumably are open-ended and cannot ever be decided finally and for eternity. In many cases, assessments and judgments may change as soon as the technological advance can considerably decrease health risks and increase reliability of the measures. In other cases, the challenge is not about balancing risk with expected utility but understanding and assessing normative issues such as human dignity, which are at the core of concerns. In the latter case, minimizing risk by technological advance will not change the outcome of assessments, as seems to be the case for reproductive cloning of humans. We will come back to these issues later (Sec. 6.3).

In the remainder of this chapter, I will draw on a recent study by the Office of Technology Assessment at the German *Bundestag*, "Monitoring the status quo and developments of pre-implantation genetic diagnosis in Germany" (TAB, www.tab-beim-bundestag.de/en/).[1]

Since its foundation, TAB has worked on different aspects of genetic engineering in several studies. The study on genome editing applied to humans was the most recent (https://www.tab-beim-bundestag.de/en/research/u30900.html; accessed 21 Sep. 2020).[2] The objective of this study was to determine the *status quo* regarding both germline therapy and somatic gene therapy via genome editing procedures. Furthermore, its task was to analyze questions regarding medical justifiability, potential risks, and ethical acceptability, the necessity of a review and revision of legal requirements in Germany, and options for regulation at the international level. The study includes an overview of the state of the art as well as chapters on more reflective issues. A recent study by the Netherlands' Rathenau Institute (van Baalen *et al.*, 2020) also provides a nice overview of the scientific facts and the social and ethical dimensions, thereby focusing on public dialogue and participation in the understanding of inclusive technology assessment (Grunwald, 2019a, Chap. 4).

6.2 Genome Editing: Turbo Genetic Engineering

Genetic engineering for intervening into organisms by introducing new genetic material has been developed since the early 1970s (Hamilton, 1972). Research has been conducted into methods of modifying the DNA of bacteria and other organisms, including plants, animals, and humans. Limitations to the applicability and efficacy of this technology resulted from the random nature of inserting DNA into the genome of the target organism. This random character

---

[1]TAB was founded in 1990 (Hennen *et al.*, 2012). Its principal is the *Bundestag's* Committee for Education, Research, and Technology Assessment, which is responsible for determining TAB's agenda. TAB's issues have to be decided *in consensus* of all parties represented in the *Bundestag*, rather than simply applying the majority rule. The various requests and topics are addressed by combining TAB's own expertise with expertise from external scientific institutions. The results are concentrated on the legislature's advisory requirements and summarized in the form of reports to be presented to the Committee or other bodies of the parliament. The results of TAB studies have led again and again to *Bundestag* debates and resolutions with direct or indirect influence on opinion formation and decision-making in the legislature.
[2]The study is available in full text in German language, including a summary in English (Albrecht *et al.*, 2020).

implies possible negative effects or unintended genetic change within the organism. Therefore, new methods were researched to allow a better targeted editing of specific sequences within a genome, in particular for gene therapies and for preventing certain genetic diseases. Genome editing based on the CRISPR-Cas9 system fulfills many of these expectations (e.g., Cong *et al.*, 2013; Adli, 2018).

Genome editing, also named genome engineering or gene editing, is the most recent variant of genetic engineering. It is used to insert, delete, replace, or modify DNA in the genome of the target organism. Its challenge is to make the desired alteration precisely at the correct position in the DNA and to avoid unintended changes. Genome editing allows the insertions to be placed much more accurately than previous approaches, in site-specific locations (Rodríguez-Rodríguez *et al.*, 2019). CRISPR-Cas9 genome editing is built on a simplified version of the bacterial CRISPR-Cas9 antiviral defense system. By bringing the Cas9 nuclease together with a synthetic guide RNA into a cell, the cell's genome can be cut at a desired location, which explains the popular notion of "genetic scissors," or "molecular scissors." This cut then allows the removal of genes or the addition of new ones in the desired place *in vivo*. Different approaches are used, e.g., so-called knock-in and knock-out mutations (Miura *et al.*, 2017). However, accuracy and precision are still a concern, e.g., related to risks of inducing genetic change beyond the targeted position in the DNA (Albrecht *et al.*, 2020).

> In other words, though CRISPR-Cas9 is the best gene manipulation technique currently available, its targeting efficiency is still inadequate (Li *et al.*, 2019).

In the meantime, newly engineered variants of the Cas9 nuclease are available that further reduce unwanted off-target activity (e.g., Akcakaya *et al.*, 2018; Chakrabarti *et al.*, 2019). Therefore, many researchers expect further decreases in unintended side effects related to off-target effects according to improved accuracy, which will extend the usage of gene therapy in many application fields.

CRISPR-Cas9 genome editing techniques have potential applications in various fields, in particular in agriculture, because of their accuracy but also because they are regarded as easy to use and relatively inexpensive. As an example from agriculture, securing global food supplies in a sustainable and environmentally-friendly manner is among the major challenges of today, in particular regarding further population increase and climate change. Traditional techniques provide improvement in some respects but also encounter limitations of applicability, efficiency, and acceptance. Genome editing provides better opportunities for progress because of its precision. According to the expectations of researchers, the risks involved in altering genomes through genome editing technology are significantly lower, compared to those associated with traditional GM crops (e.g., Zhang *et al.*, 2018). The introduction of genome editing into modern breeding programs can facilitate precise improvement for many species, at least according to the expectations of many scientists and managers of the food industry.

The field of animal enhancement (Chap. 5) for human purposes is a further huge application field (Hatada, 2017), e.g., editing of farm animals for improved food production (e.g., Yuyun *et al.*, 2016), raising questions for responsible innovation (Bruce and Bruce, 2019). Many examples of successful experiments with ideas for applications have been reported (e.g., Hatada, 2017). A recent example mentioned in the previous chapter addresses the issue of cow's milk, which can cause allergy for some people. Researchers modified the genome of a cow with the effect that a cow will then produce milk free from that allergen (Wei *et al.*, 2018; Zhaolin *et al.*, 2018). In particular, high interest is given to sheep and goats (Proudfoot *et al.*, 2015):

> Gene-edited sheep and goats provide valuable models for investigations on gene functions, improving animal breeding, producing pharmaceuticals in milk, improving animal disease resistance, recapitulating human diseases, and providing hosts for the growth of human organs (Kalds *et al.*, 2019, abstract).

After these excursions to applications of genome editing in plants and animals, we come back to CRISPR-Cas9 applications to humans. While the distinction between *somatic* and *germline* gene therapy, depending on the cells in which the DNA is modified, applies to both animals and humans (Ermak, 2015), we will focus on human germline intervention in this chapter for exploring their specific ethical challenges.

With somatic gene therapy, targeted normal cells in organs of a human body can be modified by CRISPR-Cas9 editing, for example, in the lung or the liver, to heal genetic diseases. As an example, if the therapy were used to treat cystic fibrosis, the defective gene in a lung cell would be edited in a lab. The "repaired" (edited) cell then has to be re-inserted into the patient's body. For the successful treatment of cystic fibrosis with somatic gene therapy, an enormous number of lung cells in the person's body would have to be modified genetically (van Baalen *et al.*, 2020). However, the consequences of the editing would solely affect the individual person, as long as the DNA in reproductive cells, or cells that can grow into reproductive cells, is not touched. Future generations, therefore, would not be influenced.

Many expectations for the application of CRISPR-Cas9 genome editing technology to humans therefore address healing human diseases through somatic gene therapy. One example is the development of targeted animal models for disease research (e.g., Rodriguez-Rodriguez *et al.*, 2019; Ma and Liu, 2015; Freiermuth *et al.*, 2018; Sato and Sasaki, 2018). The CRISPR-Cas9 system can also be introduced into the target cells in order to model and trace the spread of diseases and the response of the cells and organisms to infections. Introducing CRISPR-Cas9 *in vivo* makes creating more accurate models of gene functions possible.

Research in genetically edited animal models suggests that therapies based on CRISPR-Cas9 technology could have potential to treat cancer. First trials involving CRISPR-Cas9 started in 2016 by removing immune cells from people with lung cancer, then using CRISPR-Cas9 to delete the gene expressed PD-1, finally inserting the altered cells back to the same person (Su *et al.*, 2016). Other trials are on track for particular applications, among which breast cancer is mentioned repeatedly (Albrecht *et al.*, 2020). Gaining results usable in clinical therapy, however, still needs time for validation and checking possible risk (Ishii, 2017).

The main focus of using genome editing for medical purposes, however, is on diseases with a genetic cause, in particular with a monogenetic origin. Hopes have been expressed to apply CRISPR-Cas9 to heal special diseases such as beta-thalassemia, sickle cell disease, hemophilia, cystic fibrosis, Duchenne's muscular dystrophy and Huntington's disease (van Baalen *et al.*, 2020; Albrecht *et al.*,

2020; cp. for particular cases: Liang *et al.*, 2017; Ma *et al.*, 2017). These diseases affect low numbers of people but have disastrous consequences. They are an excellent target for CRISPR-Cas9 technologies because they originate in gene defects. First gene therapies based on CRISPR-Cas9 are in discussion to be approved by some national health organizations.

CRISPR-Cas9 technology also could be used for tissue engineering and in regenerative medicine and for developing novel forms of antimicrobial therapy. It may give new impetus to the concept of xenotransplantation (Tisato and Cozzi, 2012) (the transplantation of animal organs into humans), because it can help to overcome the issue of retroviruses being a major obstacle to the use of animal organs for medical purposes in humans.

A crucial difference has to be noted between germline therapy and somatic gene therapy. Germline therapy aims at intervening into the human germline (e.g., Ma *et al.*, 2017; Cussins and Lowthorp, 2018), mostly in order to prevent diseases, by repairing genetic disorders. In this field, the modified DNA *can* be passed on to future generations (cp. the case of He Jiankui discussed in Sec. 6.1), thereby raising totally different ethical issues (Sec. 6.3).

To intervene into the germline of an embryo, the embryo has to be created outside the body by *in vitro* fertilization. The CRISPR-Cas9 system can then be introduced into the nucleus of a fertilized egg cell, guided by a targeted RNA. It acts as molecular "scissors" (see above), cutting the DNA at the target position and repairing it correctly. To increase the probability that all the cells in an embryo are being modified properly, the intervention should be undertaken as early as possible in the embryo's development. After a few days, a cell can be removed from the developing blastocyst to check whether the intervention was successful and no undesirable modifications have taken place (van Baalen *et al.*, 2020; Albrecht *et al.*, 2020).

Potential benefits of modifying heritable DNA for prospective parents and their future children are expected, in particular by preventing genetically caused diseases. Mass media reports talk about far-ranging visions of the possible eradication of terrible diseases such as cystic fibrosis or Huntington's (van Baalen *et al.*, 2020). Thereby, parents with a heritable disorder could possibly be offered the option of not passing the disorder on to their children. Many of these visions have developed, from the early times of genetic engineering (Anderson, 1972). The availability of the CRISPR-Cas9 genome editing technologies fuels expectations that now there could be a real chance to fulfill at least some of these dreams. Creating HIV resistance, as was the aim of the Chinese researcher He Jiankui (Sec. 6.1), or inserting resistance against other serious infectious diseases, however, can face huge risks and challenges because, as in many other genetic interventions, a wide range of unintended consequences is possible (Albrecht *et al.*, 2020).

6.3 Ethical Issues of Human Genome Editing

The possible applications and visions mentioned show risks and ambivalences resulting in normative uncertainties, moral conflicts, and legal issues (Braun *et al.*, 2018). The possibility of failure and genetic disorder caused by germline genome editing cannot be excluded. Complex consequences including risk due to multi-gene interrelations and epigenetic effects can occur. With respect to normative uncertainty, human genome editing separates into two fields: somatic gene therapy and germline intervention. The basic difference is, simplifying a little, between healing genetically induced diseases *in vivo* in somatic therapy and preventing them in germline intervention. The latter has to be conducted in the earliest stage of embryos by editing and modifying their genome. This change will then be transferred to all children and subsequent generations of that person.

Besides medical intervention, germline intervention is highly controversial, in particular at the occasion of genetic enhancement and so-called "designer babies." This issue is significant in the international ethical debate and directly connects this chapter with the following one (Chap. 7) on human enhancement. The notion of "designer babies" goes back to preceding ethical debates on possible use, or misuse, of pre-implantation diagnostics (PID) (Savulescu, 2001; Green, 2007; Buchanan, 2011). PID allows the selection of embryos with desired properties from a larger number of embryos. This debate earned new attention after the CRISPR-Cas9 technology had become available (Knoepfler, 2015), because it could allow targeted design of particular properties of embryos or the later children and human persons. However, missing knowledge about the effects of individual gene variants and their complex interplay renders genetic enhancement a mere vision, or even speculation. Most of the human attributes and properties which are attractive to design and enhancement for some people, such as appearance and cognitive capabilities, are influenced by a very large number of genes, while genome editing is, at least to date, a powerful technology to alter single – or a small number of – genes in a targeted manner.

Somatic gene therapy is mostly still in the research or trial phase. Major fields of application are monogenic diseases and cancer therapies based on modifications of the immune system (Sec. 6.2). Current approaches are still confronted with risks, while genome editing technologies have considerably improved the perspectives of healing, e.g., certain types of cancer (Qiu, 2016; Dunbar *et al.*, 2018) and some monogenetic diseases. Compared to previous approaches in gene therapy, genome editing for somatic gene therapy does not raise fundamentally new ethical or legal questions (Albrecht *et al.*, 2020; van Baalen *et al.*, 2020). CRISPR-Cas9 genome editing has contributed to normalizing somatic gene therapy with regard to ethical and responsibility issues. As is usually the case with any new therapy, the main focus is on responsibly balancing benefits and risks. Therefore, somatic gene therapy has developed into a more or less normal field of medical practice guided by medical ethics, with responsibility, accountability, and liability configurations being well established. While this issue repeatedly gives rise to controversy and debate, its nature is well known from many preceding debates (e.g., Green, 2007; Buchanan, 2011). Therefore, I will focus on germline therapy in the next section.

6.3.1 The Consequentialist View of Germline Intervention

The ethical and legal discourse on germline therapy wavers between application scenarios full of hope and postulates for funding and supporting regulation, on the one side, and postulates of a limited moratorium or permanent ban, on the other, depending on moral positions (Baumann, 2016; Gyngell *et al.*, 2017; van Baalen *et al.*, 2020). From a helicopter's perspective, the critical and precautionary voices are by far predominant. Many ethics commissions and other advisory bodies have published statements on the ethics of germline intervention, reflecting moral controversies.[3]

The Berlin-Brandenburg Academy of Sciences and Humanities in Germany published a statement for dealing with human genome editing (Reich, 2015); it endorsed an international moratorium on interventions in the human germline. Other academies of science in Germany followed with similar statements. They did not call for a general moratorium *on research* in this field but rather for further clarification regarding potential risks and opportunities as well as concerning the normative frameworks involved. The German Ethics Council dealt with human genome editing through several events and publications (Deutscher Ethikrat, 2019). The Report of the British Nuffield Council on Bioethics (2018) reached a wide audience. The European Group on Ethics in Science and New Technologies published a first report in 2016 (EGE, 2016), while a more detailed opinion is expected to be published in 2020.

In December 2015, an International Summit on Human Genome Editing took place in Washington, D.C., discussing the ethics of germline modification. The participants agreed to support basic and clinical research under certain legal and ethical guidelines. Altering embryos to generate heritable changes in humans was regarded as irresponsible at that time (Olson, 2016). Expert groups of scientific academies consisting overwhelmingly of gene researchers and life scientists, however, are sometimes less strict. Usually, they are open to the use of genome editing in research on human embryonic development, on germline gene therapies and effects, as well as on the use of embryos for research purposes, under certain restrictions (EASAC, 2017).

In 2017, the US National Academy of Sciences, Engineering, and Medicine presented a comprehensive report on human genome editing. It includes criteria for a regulatory framework under which clinical trials for germline therapies could be permitted. The Academy, however, completely rejected addressing enhancement applications such as "designer babies" (National Academies of Sciences, Engineering, and Medicine, 2017).

[3]Thanks to the colleagues from TAB Arnold Sauter and Steffen Albrecht as well as Harald König from ITAS. I learned a lot about the issue by considering their study to the German Bundestag (Albrecht *et al.*, 2020), which includes a review of the reports of ethics commissions and advisory boards. Many thanks also to Arnold and Steffen who performed a careful review of this entire chapter.

The German Bundestag requested its Office of Technology Assessment (TAB) to deliver a comprehensive study on human genome editing, as mentioned above (Albrecht *et al.*, 2020). The results of this study include several options for the parliament and other policy actors, primarily with respect to research funding and fostering ethical and public debate. TAB repeatedly postulates not only talking about visions and speculations but rather focusing on ongoing research and the next specific stages of clinical trials and practice. According to TAB, germline interventions could be further researched in a responsible manner, under certain restrictions, in order to gain better knowledge about possible risks and to learn how to minimize them. Better knowledge would also allow public and ethical debate to be conducted in a better-informed manner, beyond speculative issues. For Germany, however, this research would require modifications to the current law on embryo protection.

The Rathenau Institute in the Netherlands has published an extensive report on human genome editing (van Baalen *et al.*, 2020), following the major objective of contributing to and fueling public debate. It includes an excellent overview of the state of the art of the technology and of the ethical debate. In addition, four scenarios were presented to demonstrate how human genome editing could develop in the coming years. The aim of providing this view of possible futures was to trigger and structure public reaction and debate in order to stimulate opinion-forming among the population, under the rationale of inclusive technology assessment (Grunwald, 2019a). The Swiss TA institution also published a report on human genome editing in 2019 (Lang *et al.*, 2019), as part of a larger assessment of different application areas of genome editing.

Reviewing these studies demonstrates that most of the ethical concerns arise from the issue of risk along questions of responsibility. Most researchers consider heritable interventions in the human germline to be irresponsible, according to the current state of the art. Their main argument is the complexity of the interaction between genetic variants and environmental factors, with the consequence that even targeted alterations of genetic variants would not guarantee the absence of risk of unintended outcomes. Possible failures are (Albrecht *et al.*, 2020):

- off-target effects: unintended genetic alterations;

- on-target effects: damage to the genome on-target due to random effects or the cell's inherent repair mechanisms;
- mosaicism: presence of different, perhaps even contradicting genetic information in the different cells of the same organism (Campbell *et al.*, 2015);
- rejection of the editing materials because of reactions of the immune system.

Each of these possible effects has to be analyzed individually with respect to possible damage or risk to the humans affected, because their shape, magnitude, and relevance depend on several factors, such as the gene to be altered, its genetic surrounding, and its embodiment into the cell.

The main ethical issue related to risks of genome editing is concern about the health and welfare of the children (Baumann, 2016; Araujo, 2017; Nuffield Council on Bioethics, 2016). Because of the possible failures of genome editing mentioned, the absence of health risks *created* by the intervention for them and further generations cannot be guaranteed. A comprehensive view must also include health risks for the mother, resulting from the different treatment steps. While the latter is a well-known challenge in medical ethics and can be tackled by the principle of informed consent, the issue of possible risks for future generations is specific to germline intervention. It is difficult, and perhaps even impossible, to gain the empirical knowledge needed to allow knowledge-based ethical reflection, because creating this knowledge would imply considering those children as objects of experiments and monitoring them throughout their lives. While this is in itself a highly sensitive issue in ethical respects, it would, in addition, take decades; too long to make the potentially desired use of certain measures possible. Therefore, animal experiments and models should contribute to gaining faster reliable pictures of possible negative effects, their shape, and their probability. At a certain stage of development of any novel and complex therapies in medicine, however, the step from experiments on animals to clinical experiments with humans must be made (Ishii, 2017). Exactly at this point, the heritability of possible adverse effects involves considerable problems because the effects of gene editing cannot be reversed later on.

Besides these individual health and welfare risks, the modification of the human genome could have negative consequences for social relationships and society at large. Problems of distributive justice could occur, in terms of the availability or affordability of the treatment among different groups of people. Repeatedly, the concern is expressed that genome editing could deeply alter the relationship between parents and children. Children with a monogenetic disease could accuse their parents because they had decided against the germline intervention, which probably would have prevented this disease. Widespread genome editing could also lead to stigmatization of people with heritable disorders. It could create public pressure on parents to apply all measures available to avoid burdening the health system or social assurance systems with children carrying those diseases and needing care.

All these potential effects have been discussed for decades as a possibility at the occasion of prenatal diagnostics (Green, 2007), in particular in the context of disabled persons (Wolbring, 2006). According to the experience of recent decades, the occurrence of real effects is unfortunately more than mere speculation. The availability of prenatal tests has changed values and behavior. In large parts of the population, prenatal tests have turned from a voluntary or optional measure to a standard; not by law but by the factual development of attitudes and customs. In the case of a Down syndrome diagnosis, termination of pregnancy has become a standard reaction in many countries, making persons with Down syndrome possibly less accepted in society (Grunwald, 2019a).[4] If there were fewer people living with a particular genetic disorder because of prevention by genome editing, the awareness of, the familiarity with and the social acceptance of people with this disorder could decline (Nuffield Council on Bioethics, 2018, 84, according to van Baalen *et al.*, 2020). Independent of the arrival of "designer babies," elements of a liberal eugenics (Habermas, 2005) can already be witnessed.

The health and social risks mentioned so far are all elements of the consequentialist paradigm (Sec. 3.3.2), as risks generally are.

---

[4]As was reported recently ("Iceland close to becoming first country where no Down's syndrome children are born." *Independent.* www.independent.co.uk/life-style/ health-and-families/iceland-downs-syndrome-no-children-born-first-country- world-screening-a7895996.html; accessed 21 Sep. 2020) no more children with Down syndrome are born in Iceland, due to the effect mentioned (thanks to Julie Cook for mentioning this example).

They are concerns about possible effects in the future, either social effects in a society with more and more genome editing applications, or individual health effects for the children, or later generations. The latter could be considerably reduced by further research, as is expected, e.g., by researchers and scientific academies. Decrease in risk and increase of reliable knowledge directly affects the responsibility of those interventions, reminding us of the EEE model of responsibility (Sec. 3.3.1). The challenge remains to assess *when* the time will be ripe for giving approval to human genome editing in certain areas, i.e., whether the consequential knowledge available suffices to draw sound conclusions. In some fields, this might indeed be a question of time, while in other cases uncertainty and risk might persist.

For social risks, however, things look different. Concerns of unfair and unequal access to genome editing technologies, of a possibly emerging pressure on parents to use genome editing to prevent monogenetic diseases, or of future competition with respect to creating the best-enhanced designer babies all follow the structure of slippery-slope arguments. By means of these concerns, a specific change of human behavior is presupposed: adaptation to the technologies and an erosion of values according to that adaptation. Thus, slippery-slope arguments are frequently implicitly or explicitly built on the idea of technology determinism (Grunwald, 2019a). They do not trust humans, or human societies, to uphold and maintain values and moral standards in cases of temptation by new technology. This position, therefore, is not based on normative assessment due to ethical standards but corresponds to a specific view of humans and society (Walton, 2017). This should be debated instead of blindly followed; otherwise self-fulfilling effects of unreflected adaptation could be the consequence (Grunwald, 2019a).

A second issue with social concerns, e.g., of misuse and distributive justice, consists in their very nature. Misuse and injustice are not properties of the respective technology but consequences of human action (misuse) or deficient organization of distribution and access (justice). Thus, these concerns are not arguments against human genome editing *as technology* but rather address their social embodiment (Foladori, 2008). Therefore, concerns of this type can be converted into recommendations and postulates for, e.g., better regulation to prevent misuse, or for implementing justice-friendly political and economic boundary conditions. They do not address technology *as such* but the way and how technology is developed, adopted by, and embodied in society.

By summarizing the consequential view of human genome editing, it becomes clear that balanced and differentiated analysis and weighting are needed. In ethical respects the resulting arguments are "weak" in a twofold sense: (1) their validity depends on more or less uncertain assumptions of how the future will develop in particular respects. In many fields of NEST the consequential knowledge is so speculative that it has even been called into doubt whether this approach makes sense at all (Nordmann, 2007a). (2) Arguments based on the possible manifestation of unintended side effects in the future are not arguments against the new technology *per se*. Rather, they can be converted into arguments that something must be done politically and socially, and perhaps also technically, to limit, prevent, or compensate these unintended consequences. Both types of ethical concern are part of a precautionary and reflexive approach to the future. They aim at detecting unintended effects as early as possible and establishing countermeasures in order to contribute to maximizing gains and to minimizing risk. They belong to reflected preparations for the advent of new technology, as part of a comprehensive technology assessment with its cognitive interest of:

> [...] supporting, strengthening and enhancing reflexivity in all epistemic and social fields of reasoning and decision-making on shaping the scientific and technological advance, on the usage of its outcomes and on dealing with the consequences to present and future society (Grunwald, 2019a).

Thus, the ethical arguments mentioned above are an important part of proactive and prospective reflection and a highly needed contribution to responsible research and innovation (Owen *et al.*, 2013; von Schomberg and Hankins, 2019). Nevertheless, I denote these arguments as "weak" in ethical respects because of their hypothetical and conditional nature, which is related to their consequential origin. Things might be different in future, and then the arguments would be rendered invalid or irrelevant; in many cases, this would even be desired. If, for example, ethical concerns about unfairness and unequal distribution of CRISPR-Cas9 technology motivated policymakers to establish regulation preventing these developments, then the validity of the ethical concerns would disappear – which would be exactly the best-case effect wanted.

6.3.2 Deontological Arguments

If the arguments mentioned above are "weak," what could be said about "strong" arguments in ethical respects? Strong arguments should remain valid in a case where all the weak arguments disappear, due to complete avoidance of failure and risk, because of achieving social justice and inclusion, and so forth. Would genome editing be fully approved in this hypothetical case? "Strong" arguments must build on ideally *logic or ethical implications* of genome editing technology that *necessarily* will appear. Strong arguments in this sense would be categorical ones. They must not address contingent unintended side effects dependent on uncertain future developments but rather hint at implications of their application. This would be the case for *deontological* arguments operating, e.g., with violations of a certain moral status assigned to humans, or with violations of fundamental human rights. Indeed, some ethical arguments claimed to be "strong" on this understanding have been proposed in the field of human genome editing.

A famous example is Jürgen Habermas' argument against genetic interventions into the human germline (Habermas, 2005). According to his line of thought, an embryo as an emerging person made subject to such an intervention during its embryonic stage would be instrumentalized without any opportunity to either agree, to defend themselves, or to free themselves of this instrumentalization in later life. The determination of the genome edited at the stage of an embryo remains during the entire lifetime of the human person. Habermas concludes that this person would not have any chance to become the full author of their own biography but would be irreversibly instrumentalized due to the interests of others, e.g., their parents – or even by the state in totalitarian regimes. According to his argument, this configuration would affect and finally undermine the *conditions of morality* in general. The core of this argument is that, in this case, germline editing procedures would be applied to humans (1) being unable to give their consent and (2) being unable to reverse these interventions in later life. If competent adult persons, after receiving information about possible health or other risks and after having provided their informed consent, would undergo such an enhancement, then this argument would not apply.

The Habermas' argument against any intervention into the human germline was and still is controversial (e.g., Morar, 2015). It is rooted in continental philosophy, in particular in the philosophy of Immanuel Kant and his view on humans as autonomous beings. His understanding of human dignity, which, for example, became part of the German constitution (*Grundgesetz*), is deeply related to the issue of autonomy. Autonomy would indeed be affected by germline intervention without the possibility of informed consent and without the chance of a later removal. Consequentialist utilitarian reasoning, however, takes a different perspective and usually argues less strictly and rather opens up opportunities for balancing and weighing benefits and risks.

A further argumentation claiming to be strong in the sense mentioned is based on the assignment of moral status to embryos. National regulations and cultural traditions diverge with respect to this moral status. To put it in the extreme: is the embryo an emerging human person under the protection of human rights, or is it a kind of natural biochemical entity, which could be regarded as a simple thing open to any human intervention? If the embryo is regarded as an emerging human, which is, e.g., a widespread position of some confessions of the monotheist religions (e.g., the Roman Catholic Church), no manipulation would be morally allowed at all. The corresponding secular philosophical argument, e.g., based on Kant, however, can be challenged, e.g., by looking at disastrous monogenetic diseases. In this case, authors argue that preventing such disease by genome editing would increase the autonomy of that emerging person because it would open up better opportunities for an autonomous life. Followers of Kant would respond that this indeed could be the case but that the person affected should be able to decide what is good for themselves. This issue obviously extends to the question of whether and to what extent parents should be allowed to take the respective decisions.

In these deontological debates, terms already mentioned such as "the right to an open future," "identity," "autonomy," and "human dignity" are often used (van Baalen *et al.*, 2020). Although they are important concepts and hint at ethical principles, they often do not clarify the challenges under consideration but rather make the major questions *invisible* because of their appellative character and diffuse meaning. Almost nobody would oppose defending human dignity and autonomy, but different persons often presuppose different meanings, enabling them to draw diverging conclusions. The Universal Declaration on the Human Genome and Human Rights (UNESCO, 1997) states that the human genome underlies the fundamental unity of all members of the human family (Article 1). The explanatory memorandum to the Dutch Embryo Act raised the question of whether respect for an individual's dignity meant that a person had the right to inherit a genetic pattern that has not been altered through targeted human intervention, or whether it means that modification of a person's DNA to prevent them from inheriting a heritable disorder actually accords with that principle (Kamerstukken II, 2000/2001, 27 423, no. 3, p. 45, according to van Baalen *et al.*, 2020). The Health Council of the Netherlands took the view that removing a serious inhibiting disease benefits a person's dignity and their right to an open future and individual autonomy (COGEM & Gezondheidsraad, 2017).

Independent of these considerations, the question of the moral status of an embryo remains. Any answer will have dramatic consequences because genome edited embryos always have to be tested to establish whether the intervention had the desired consequences. If not, these embryos would have to be destroyed or annihilated (for this, completely value-neutral formulations probably are not available). This would not be a problem in cases of regarding them as mere biochemical entities. However, it would be a huge problem if they were regarded as carriers of a specific moral status putting them under the protection of human rights, or a regulation dedicated to protecting embryos (as is valid in Germany, for example, cp. Albrecht *et al.*, 2020). In fact, there is a huge spectrum of different opinions about the necessity to protect embryos depending on the moral status assigned to them. The opinions can be classified along four different angles on the issue (in the German language these are known as SKIP arguments) (Damschen and Schönecker, 2003):

- membership of the biological species *homo sapiens*;

- continuity of development from the earliest stage of embryos to the stage of adult persons without a sharp boundary in between;
- identity of the organism in history over time but also, in biological terms, according to its genome;
- potentiality of the embryo already at the one-cell stage to develop to an adult person.

These perspectives, which might be abbreviated as SCIP (species – continuity – identity – potentiality),[5] are all debated in a controversial manner. No consensus is expected. That means that different national regulations and practices will persist, but so will normative uncertainty. Concerning the years to come, scientific and medical advances will have to go hand in hand with ethical reflection to allow a stepwise and responsible approach to the future.

6.4 Responsibility Challenges of Germline Intervention

Reflecting the ethical issues described above in light of the EEE model of responsibility (Sec. 3.3.1), challenges to responsibility assignments in the field of human genome editing emerge. They concern different perspectives covering specific actors and people affected, such as physicians and their professional organizations, researchers and scientists, medical ethicists, patients and their relatives, the health system including care personnel, technical assistants, health insurers, and society at large. However, while this enumeration sounds a bit traditional, we will see that it has to be extended, crucially changing the responsibility configuration for germline intervention.

In the field of germline intervention, a new entity enters the responsibility configuration: children and adults as products or outcomes of reproductive technologies involving germline genome editing. This is a strange entity compared to the actors usually involved, because it forces us to consider responsibilities with respect to persons emerging years and decades after the intervention. The "patients" of genome editing differ from familiar patients

---

[5]Thanks to Steffen Albrecht for this proposal.

(Wippermann and Campos, 2016). This configuration, therefore, creates a completely different situation with respect to normative uncertainty and resulting ambiguities in assigning and distributing responsibilities. Questions such as whether it is responsible or irresponsible to prevent a disease by genome editing, or to enrich an embryo with some desired abilities arise, as well as the question of the criteria and processes of assessment and decision-making. This challenge will be investigated in-depth in the remainder of this section, according to the EEE model of responsibility. While its ethical dimension was explored in the preceding section, I will focus here on the epistemological and empirical dimension.

Epistemological dimension

The epistemological dimension reflects the knowledge required for responsibility considerations and its epistemic quality. In the case of germline interventions, this knowledge should comprise the consequences of the intervention for all parties involved or affected. These include, first, the couple (or a single) wishing to become the parents, in particular the mother, the physicians, researchers, and technicians and possible further actors involved at the time of the intervention. Second, and decisively, the emerging human person in their later biography as well as their potential progeny have to be considered. There is wide consensus among scientists that:

> [...] gene editing techniques are not sufficiently safe or effective to be used on human reproductive cell lines. Evidence for the safety and effectiveness of this technology can only be obtained through basic and preclinical research (Li *et al.*, 2019).

Even authors calling for permission for the genetic modification of embryos do not deny the complexity and risks or the need for further research to gain better knowledge, in particular in order to minimize risk and possible adverse effects (van Baalen *et al.*, 2020). This knowledge, however, must not be gained directly by clinical experiments and monitoring their outcomes. Instead, it should be available *before* starting clinical experiments, as far as possible. Otherwise, these would be irresponsible, as was discussed at the occasion of the experiments made by He Jiankui (Sec. 6.1) but also repeatedly since the 1980s. Even where they were regarded as responsible under certain restrictions, the knowledge required would be useful only to a limited extent, because monitoring the consequences would take years and perhaps decades. Slowing down human gene editing too drastically, however, could also not be a responsible approach, because the possible positive effects, e.g., healing serious diseases, would have to be postponed longer than necessary. In a sense, this is a dilemma situation: creating knowledge to enable responsible germline intervention is possible only by violating responsibility requirements. Therefore, the knowledge required must be gained by other approaches, as far as possible: by animal experiments (Chap. 5), modeling, and preclinical research.

At some stage of development and knowledge, however, the step to clinical research must be taken, involving specific responsibility challenges. While this step is familiar to the development and implementation of new therapies in many other fields, the case of germline intervention is special because of its future dimension, reaching beyond the persons directly affected. Even if the probability might be extremely low, it could happen that the person influenced by germline editing would be healthy but that their descendants could be confronted with problems because of the complexity of inheriting and combining DNA. This epistemological issue further increases the requirements of and challenges to responsible genome editing. It means moving within the dilemma situation mentioned above. While fully observing strict ethical standards would prevent clinical trials, lowering of the standards is irresponsible for other reasons. The only way out of this dilemma is a stepwise approach, accompanied by careful monitoring of all the experiences so far and deciding on the next step, accompanied by ethical reflection, and so forth.

Empirical dimension

As indicated at the top of this section, the special and particularly challenging property of the responsibility configuration in our case is that a new entity enters the responsibility configuration: children and adults as products or outcomes of reproductive technologies involving germline genome editing. Their embryos are, in a sense, the major objects of germline intervention. Do they have a place in responsibility assignments? What place could this be, and what would the corresponding responsibility distribution look like?

First of all, we have to fully understand the deep asymmetry in this case. The embryo is the main object of the intervention: the germline to be edited is *its* germline. This edited germline will accompany the emerging human person throughout its entire life, which hopefully will be "better" than it would have been without intervention, but without full guarantee of this. The emerging human person – the embryo – does not have any *subject* role in the process but is only *object* to intervention into its germline. Reminding us of the introduction of responsibility (Sec. 3.3.1), the embryo is indeed a mere object, which has to bear all of the consequences of the germline intervention in its later life. It will always remain bound to the intentional decisions on its germline made by others, mostly its parents, and be unable to remove or reverse the changes made. This deep asymmetry cannot be addressed in terms of familiar risks and benefits. By making far-reaching decisions about their children's genes, "human reproduction will be transformed into the production of humans" (van Baalen *et al.*, 2020, 43). This is what Jürgen Habermas (Sec. 6.3) denoted when he said that humans with an edited germline would not have a chance to become full authors of their own biography but would remain products of the decisions made by others.

Other "entities" are also affected, which is often neglected. In order to intervene into the germline of an embryo, several embryos have to be created by *in vitro* fertilization. Those embryos where the intervention was not successful will be destroyed. Therefore, embryo destruction belongs to successful germline interventions. Depending on the assignment of moral status to an embryo (cp. previous section), this also has to be taken into account while reasoning on responsibility (Albrecht *et al.*, 2020).

Because of the time between the initiating act – the germline intervention – and possible consequences years or decades later, responsibilities have to be considered over time and raise the issue of solidarity over time (Mulvihill *et al.*, 2017). At the time of the decision-making and intervention, the responsibility configuration is quite clear in the empirical dimension: the parents in cooperation with the medical team are responsible for their actions and decisions. While this sounds quite normal, it is not. An "invisible" or "absent" entity is part of the game: the not-yet-existing human person (potentially) emerging from the embryo to be edited. Obviously, it is not possible to involve the emerging person in the decision-making process and to get its informed consent. Therefore, a (literally) paternalistic decision-making regime is unavoidable. In many other fields, this is established and legitimate, e.g., when parents decide about health issues relating to their children, such as vaccinations. Germline editing is different because it affects the genome of the embryo, which is possibly interwoven with human identity and personality (Sec. 6.5). Therefore, we cannot avoid paternalism but have to ask for a conscious "responsible paternalism" under these requirements.

This paternalism primarily concerns the decision-making process: What are the goals and targets of the intervention; which gene is to be knocked out or replaced; should the intervention be undertaken at all; what could the risks be; what alternatives are available with what implications; and how could a decision be made in case of diverging or counteracting arguments? As a general goal, we can assume: giving the future children/adults the opportunity to have a better life than they would have had without germline modification. This could be realized, e.g., by genome editing preventing a serious genetic disease which causes suffering, low quality of life, short life span, and so on. While this story, which is usually told by promoters of germline editing, sounds very plausible and appealing, the situation in other proposed interventions is more complex and partially ambivalent.

In approaches to genetic enhancement, the issue is not to prevent disease but to improve certain abilities of the emerging persons (Wolbring, 2015; Campbell and Stramondo, 2017), arguing in the same way that enhanced persons will have a better life because of their enhanced capabilities. In both cases, the core notion is the small word "better." It sounds, at first, quite clear and plausible. Everybody prefers to be "better" than "worse." However, can consensus about the meaning of "better" be presupposed between the decision-makers and the later human person? Is it allowed and responsible to transfer an agreement among the parents to the emerging person? In cases of preventing serious disease this seems plausible, but in cases of enhancement, however, this might not be so clear. Priorities, lifestyles, and values may change over time, in particular over decades. In addition, it should be considered that improvements, betterments, and enhancements in a particular respect can often show negative or at least ambivalent implications in others (Sec. 5.3.1, Sec. 7.2).

Therefore, the core of the responsibility configuration consists of the question of the legitimacy of transferring the criteria of "better" from the decision-making persons to the person emerging out of the edited embryo, not forgetting that that person has to live with this "betterment" throughout their entire life. This is particularly challenging because there will be a large grey zone between extremely plausible situations and questionable ones. Determining the boundary between responsible and irresponsible determinations of future persons would therefore be a major issue if genome editing was to become a real option, beyond research.

A further responsibility challenge will be met as soon as adverse effects and health problems directly related to the gene editing occur in the lifetime of the emerging person. These would have a clear origin: the intentions and decisions of parents and/or the actions of the medical team. If a failure of the medical team was the origin, then the familiar regulations of liability would apply. However, if the desire of the parents was the origin, they could be made responsible and liable, e.g., with respect to incurring the costs of any medical care needed. In any case, the affected person must not be made responsible because they were not involved in creating the problem.

In thought experiments, another possible problem has also been identified. Parents will focus on the future properties of their child, e.g., the absence of a severe monogenetic disease, intended HIV resistance (as was the intention of the parents of the Chinese twins Lulu and Nana), or perhaps some abilities or properties related to appearance, or cognitive or sporting capabilities in cases of enhancement. Assume that the genome editing process would lead to a human child with different properties than the parents wanted them to have. They had paid for the intervention with specific expectations, which then would not be fulfilled. The "product" delivered would not correspond to the product "ordered." Would such an outcome be a reason to refuse payment or to hold someone liable, comparable to the liability of companies for their products? And, what would happen if those parents would not accept that child in their family? Obviously, these are mere speculations. But they show the dramatically wide field of possibilities and huge normative uncertainties which can be opened up by germline intervention.

Thus, it turns out that the most challenging responsibility issue is making the basic decision to intervene into the germline. Related responsibility considerations have to build on moral positions in this field. It seems impossible to reach full or wide-ranging consensus among the various societal groups in this respect in the next few years. Perhaps this will always remain a contested issue because of deep-lying questions such as the moral status of the embryo and various religious pre-occupations on the nature of human life (e.g., Al-Balas *et al.*, 2019) but also because of the likely remaining uncertainties about risks and adverse effects for the edited persons and their descendants.

In this situation, responsibility can only be assured by means of well-reflected *governance* of the field (Li *et al.*, 2019) and by conducting respective processes properly. After the experiments by He Jiankui, which were regarded as irresponsible (Sec. 6.1), many activities were initiated to review the field's responsible governance, based on previous statements and proposals. Many commissions and advisory bodies had already published their perspectives and recommendations, such as the European Group on Ethics in Science and New Technologies (EGE, 2016), the National Ethics Council in Germany (Deutscher Ethikrat, 2019), the U.S. National Academies of Sciences, Engineering, and Medicine (2017), the Health Council of the Netherlands, and the Chinese Academy of Sciences (some of these reports were mentioned in Sec. 6.3). All of these position papers and recommendations speak about a distributed responsibility among science, regulatory bodies, the medical system and patients. Some of them are rather strict and postulate a complete ban on germline intervention, including further research (Baylis *et al.*, 2019). Most of the recommendations, however, follow a precautionary approach. They distinguish between preclinical basic research, which is regarded as being allowed and is sometimes explicitly desired, if done in a responsible manner, and clinical research. The latter is usually regarded as irresponsible, according to the current state of the art, which has motivated postulates of a moratorium (Chneiweiss *et al.*, 2017; Daley *et al.*, 2019). All of the recommendations give strong emphasis to public dialogue on the ethical standards and responsibility distributions, which should be conducted in an inclusive manner rather than being restricted to listening to experts (Jasanoff *et al.*, 2015).

Due to differing cultural and historical backgrounds, national regulation differs from country to country, e.g., between Germany and UK on the issue of embryo protection, while many countries do not have any regulation on these issues at all. Probably, this situation will persist. However, some conditions should be fulfilled:

- A set of minimal standards of responsibility for gene editing should be accepted and observed worldwide, based on the International Declaration of Human Rights approved by the United Nations in 1948. This declaration can be regarded as a kind of "constitution" of the global human community.
- Regarding different national (and regional) regulations, gene-editing tourism of parents travelling to countries with a more permissive legislation will happen, as is well known from other cases, such as women accessing termination of pregnancy, or researchers doing their animal experiments in countries with low animal protection standards (Chatfield and Morton, 2018; Rath, 2018). Therefore, rules for this kind of tourism should be established at the global level.
- The overall governance of human germline gene editing should be transnational and global at the level of meeting consensual ethical standards, while it may differ in detail across nation states (Rosemann *et al.*, 2019).
- Transcultural dialogue is needed between Western countries, China and other Asian traditions, and other positions, in particular from the Islamic world and from the Global South, in order to strengthen mutual understanding. A global observatory could support this process (Jasanoff and Hurlbut, 2018).

Regulations and laws have to be developed and continuously improved in order to meet the rapid development of genome editing technologies in various application fields. Accordingly, existing technical and ethical guidelines must be refined and adapted. Capacity-building with respect to the ethical review capacity and interdisciplinary cooperation at all levels should be strengthened, including researchers, medical practitioners, interested parents, and civil society (Li *et al.*, 2019).

6.5   Genome Editing: Editing Humans?

Considering the debates on the significance of the human genome for human identity and of its role regarding properties and abilities of individuals, the question arises whether editing human genomes implies editing humans (cp. van Baalen *et al.*, 2020, 43). Following the idea of genetic determinism, which was dominant in the 1990s, this question would have to be answered "Yes." If humans were determined by their genome with respect to some issues that shape their identity, then editing genomes would directly be an editing of human persons. The changes made at the stage of embryos would directly impact and determine the emerging human persons. Genetic determinism:

> [...] is the belief that human behavior is directly controlled by an individual's genes [...], generally at the expense of the role of the environment, whether in embryonic development or in learning ("Biological determinism." *Wikipedia*. https://en.wikipedia.org/wiki/ Biological_determinism; accessed 21 Sep. 2020).

Biologist Richard Dawkins acted as a major representative and public disseminator of this conviction, influencing public opinion at large (Dawkins, 1989). In the meantime, however, genetic determinism was challenged by several findings, in particular on the development of twins, which called the determinist nature of genes into doubt. Currently, most scientists believe that the dichotomy between nature and nurture (Ridley, 2003) does not make sense. According to the state of the art, genes are expressed within an environment, may have variable expressions in different environments and are continuously influenced by their environments. Many mechanisms of the mutual interdependence of genes and environment have been detected in the field of epigenetics (Moore, 2015).

In spite of the rejection of genetic determinism, the question remains how far editing human genomes also could be editing humans. Without any doubt, genome editing changes the genetic inventory of future persons and thereby impacts these persons. The question is whether and how far the person's identity will be touched. How far does HIV resistance or having or not having a gene responsible for a monogenetic disease determine personal identity?

While issues of this type show considerable potential for debate, this would be even larger if genetic enhancement ("designer babies") became a reality, because in those cases, influencing the identity of the emerging persons belongs to the intention of the intervention.

European legislation prohibits clinical tests "which result in modifications to the subject's germ line genetic identity." (Regulation [EU] 536/2014, Article 90). But what does the term "genetic identity" mean? It can be understood, even if this probably was not intended, to assume genetic determinism, in that human identity is entirely embodied in the DNA (van Baalen *et al.*, 2020). Independent from believing in genetic determinism, the role the genome plays in our thinking about its significance and role in human beings must be reflected, without reducing human identity to "genetic identity." In the notion of a "psychosocial" identity, emphasis is given to how individuals perceive and shape their identity within their social context (Nuffield Council on Bioethics, 2018). From this perspective, the impact of genome editing on personal identity primarily depends on how the edited person perceives and shapes their identity. On this understanding, the fact that a person knows they have been edited according to the intentions of their parents (or others), could be more influential on their identity than the biological consequences of the editing process itself.

These brief reflections, while remaining at the surface level, give some indication that human identity is a complex and contested construct. It should not be reduced to biological and chemical DNA configurations in the narrow understanding of a "genetic identity." They also show that the interface between human *life* and *technology* is in deep movement, according to the main hypothesis of this book. This will give rise to further reflections on the future of life (Chap. 9).

Chapter 7

Human Enhancement

A controversial international debate about enhancing human beings through new technology has been ongoing for almost twenty years. Building on previous considerations of improving humans by genetic engineering or eugenics, this more recent wave occurred after a publication by the National Science Foundation (NSF) (Roco and Bainbridge, 2002). The *debate on human enhancement* enriched and fueled intellectual reasoning about the future of human beings by adding further ethical, anthropological, and philosophical aspects, in particular in the fields of post- and transhumanism (Hurlbut and Tirosh-Samuelson, 2016). This chapter introduces the central ideas of these underlying enhancement stories (Sec. 7.1), clarifies the semantics of technologically improving humans (Sec. 7.2), summarizes the major concerns raised in the ethical debate so far (Sec. 7.3.), considers possible political dimensions underlying this (Sec. 7.4), addresses responsibility issues (Sec. 7.5) and, finally, investigates the changing relations between humans and technology in the course of human enhancement (Sec. 7.6).

7.1 Technologies for Human Enhancement

Human enhancement has been an issue in human culture and history since the earliest times (Sec. 7.1.3). Technology, however, was and still

Living Technology: Philosophy and Ethics at the Crossroads Between Life and Technology
Armin Grunwald
Copyright © 2021 Jenny Stanford Publishing Pte. Ltd.
ISBN 978-981-4877-70-1 (Hardcover), 978-1-003-14711-4 (eBook)
www.jennystanford.com

is used mostly as *external* support for humans, without intervening into their nature, except for medical purposes. Improving humans by technology is a rather new issue, fueled by the NSF report (Roco and Bainbridge, 2002) (Sec. 7.1.1). In this report, several perspectives for enhancing humans were proposed, which still frame large parts of both research and public communication (Sec. 7.1.2).

7.1.1 Converging Technologies for Improving Human Performance

The NSF report *Converging Technologies for Improving Human Performance* includes highly visionary, ambitious and partially provocative perspectives on human enhancement, which it claims may or even will become real in the next few decades. According to this report, the NBIC converging technologies of nanotechnology, biotechnology, information and communication technology, and cognitive and brain science offer far-reaching perspectives for perceiving even the human body and mind to be formable and for improving them through precisely-targeted technical measures. Human enhancement is understood here to be *technology-based* enhancement, initially at the level of individual abilities but also addressing the level of society:

> Rapid advances in convergent technologies have the potential to enhance both human performance and the nation's productivity. Examples of payoff will include improving work efficiency and learning, enhancing individual sensory and cognitive capacities, revolutionary changes in healthcare, improving both individual and group efficiency, highly effective communication techniques including brain to brain interaction, perfecting human–machine interfaces including neuromorphic engineering for industrial and personal use, enhancing human capabilities for defence purposes, reaching sustainable development using NBIC tools, and ameliorating the physical and cognitive decline that is common to the aging mind (Roco and Bainbridge, 2002, 1).

Nanotechnology is key to these expectations because the technologies mentioned could meet, converge, and even unify at the level of atoms and molecules. Each physical process, but also each process in living organisms, has a grounding in the world of atoms. Therefore, if it were possible to control matter at the level of atoms and molecules, promoters of converging technologies are optimistic that all the processes at more macroscopic levels could then also be controlled (this "atomic reductionism" is, however, highly contested in the philosophy of science, cp. Janich, 2006):

> Nanotechnology is the means for manipulating the environment at the molecular level. Through modern biotechnology, humans endeavor to direct their own evolution. Information technology and cognitive science are both exploring ways to increase the speed and range of information a person can assess, whether by artificial computing or expanding cognitive capacity (Williams and Frankel, 2006, 8).

Several alternative interpretations of the meaning of convergence have been proposed (for more details, see Coenen, 2008; Wolbring, 2008b). They comprise the following:

- *Technological convergence*: An evolutionary convergence of the technical fields named above, in the sense that an overarching line of technology is formed that replaces the previously separate fields and in which the individual fields largely merge.
- *Convergence of the sciences and engineering*: The convergence of the disciplines of the natural sciences with those of the engineering sciences in the sense of *technoscience* (Asdal *et al.*, 2007), which is the foundation of the European answer to the NSF report's position (EC, 2004).
- *Unification of the sciences*: Development toward a new unity of the sciences on the basis of nanotechnology, referring to Leonardo da Vinci by expecting a "new Renaissance" (Roco and Bainbridge, 2002, 2). This idea of convergence links with sociobiological programs and with visions of a new unity of science (Bunge, 2003; Wilson, 1999).
- *Generator of issues for the agenda of science*: As a concept of research policy, the idea of convergence promotes the further development of the scientific disciplines involved, for their agenda-setting, and a motivation for funding agencies.

In this volume, atomic reductionism is assumed as the guiding motif behind many of the visions and expectations related to the converging technologies, as expressed in the slogan "shaping the world atom by atom" (NNI, 1999). In this line of thought, convergence means enabling a new ground for determining common roots of the scientific disciplines involved, namely that they all operate and work at the level of atoms and molecules. The traditional scientific disciplines would converge in these roots; in analogy to plants, they can be viewed as different shoots that have developed out of common roots. Operating with atoms and molecules would thus become an engineering version of a "theory of everything." This would not be a theory of physical fundamentals that provides explanations at an abstract level. It would be a practical theory of operation and manipulation in order to construct any matter in a targeted fashion. Ultimately, nanotechnology – as such a theory of everything in this line of thought – would also deliver instructions and know-how to build and operate a molecular assembler (Drexler, 1986), or similar constructions. From this perspective, the spheres of the living and the social are supposed to be explained and made the object of technical manipulation starting from their atomic basis, in a radicalized version of nineteenth-century physical reductionism:

> Science can now understand the ways in which atoms form complex molecules, and these in turn aggregate according to common fundamental principles to form both organic and inorganic structures. [...] The same principles will allow us to understand and when desirable to control the behavior both of complex microsystems [...] and macrosystems such as human metabolism and transportation vehicles (Roco and Bainbridge, 2002, 2).

This new science would signify the triumph of atomic reductionism. Nanotechnology, as the science of working technically with atoms and molecules, would provide the necessary key competence of this new unified science, which would reconcile discovery, cognition, and engineering. At the same time, it would allow humans to build not only amazing machines such as molecular assemblers (Drexler, 1986) but also to ultimately create themselves anew technologically, thus modifying ourselves according to our ends, going beyond mere enhancement:

> When God fashioned man and woman he called his creation very good. Transhumanists say that, by manipulating our bodies with microscopic tools, we can do better (Hook, 2004; see also Dupuy, 2007).

If this program could be implemented, then even the categorical difference between the living and the nonliving would disappear, at least with regard to the capacity for matter to be manipulated at the atomic level. The following quotes exemplify the optimistic belief in technological progress expressed in the NSF report. They are not only supposed to be milestones on the road to such a new society and to a new human but also to provide an impression of what is understood by technical enhancement of humans:

> Fast, broad-bandwidth interfaces directly between the human brain and machines will transform work in factories, control of automobiles, ensure superiority of military vehicles, and enable new sports, art forms and modes of interaction between people.

> The human body will be more durable, healthy, energetic, easier to repair, and resistant to many kinds of stress, biological threat and aging process (Roco and Bainbridge, 2002, 4ff.).

> Convergences of many technologies will enhance the performance of human warfighters and defenders (Roco and Bainbridge, 2002, 287).

These quotes claim to describe more than simply fictive stories. Roadmaps and milestones are supposed to provide the link between today's research and policy and visionary futures (Roco and Bainbridge, 2002, 4ff.). In this sense, the visions claim to formulate not *envisaged possible* futures but to be anchored in today's scientific activity by being part of the current agenda and its milestones (Grunwald, 2007; Coenen, 2008). Consequently, the language chosen comes close to telling us about existing roadmaps of future developments, and not just about visions or possibilities.

The NSF report was published almost twenty years ago. In some respects it has shown itself to be false in the meantime, in particular when concrete years passed by without certain milestones having been reached. For example, the capability to download the consciousness of humans to a computer, which was predicted for 2025, is probably still as far off as it was twenty years ago. However, such stories do not devalue the report, because it attracted worldwide attention (e.g., EC, 2004; Coenen *et al.*, 2009) and has left traces that are still powerful today. Some of the stories have migrated to other technology debates. While the NBIC stories themselves were interpreted as a kind of legacy of the visionary nanotech debate (e.g., Drexler, 1986; NNI, 1999) (cp. Wolbring, 2008b), they found successors e.g., in the fields of synthetic biology (see Chap. 4), digitalization and artificial intelligence (Chap. 8). In this chapter, the focus is on the field of human enhancement as the direct target of the NSF report.

7.1.2 Visions of Human Enhancement

The NSF report presents many ideas and utopias for improving human performance, most of them focusing on the capabilities of individual humans rather than at the level of society. The envisaged enhancements can be subsumed under three headings: (1) enhancing human capabilities by expanding cognitive or nervous functions (*Neuro-enhancement*); (2) adding completely new functions to humans (*Extension*); and (3) lengthening the life span of humans (*Immortality*).

(1) *Neuro-enhancement*

Neuro-enhancement is to be among the key application fields of converging technologies (Roco and Bainbridge, 2002; Farah *et al.*, 2004). Linking digital technologies with the nervous system is at its core. Neuro-enhancements refer to improvements of human capabilities resulting from implants or medications either connected to the nervous system or to the brain, or that act upon them. Any human capability which can be considered analogously to the performance of a machine in engineering language, can be regarded as a candidate to be improved by technical means. In the case of neuro-enhancement, humans are modeled as information-processing beings such as robots. From this perspective, it is possible to distinguish technical enhancements for taking up data from the environment (sensors), technical enhancements of the brain's capacity to process and store data, and technical enhancements for controlling external motor systems.

To realize neuro-enhancement, it is necessary to transform electrical impulses from contacts or wires linked to digital technologies and information systems into impulses perceived and understood by the nervous system, and vice versa (e.g., Abbott, 2006). For example, a successful interconnection between an artificial limb (e.g., hand or leg prosthesis) and the nervous system would make it possible to compensate for the loss of a natural limb. Animal experiments are used for preparing applications to humans (Chap. 5). While this sort of technical limb reconstruction would open up possibilities for restoring lost human capabilities, it can also be used to expand conventional body functions beyond the natural level.

Neuro-implants are developed to restore, replace, and perhaps enhance human sensors or organs. Often, the starting point is the motivation to compensate technically for a *loss* of sensory function (e.g., of sight or hearing). As a result of advances in nanoinformatics, such as miniaturization or the increased capacity of implants to take up and process data, the spatial dimensions and performance of the neuro-implants may approach those of natural systems. A visual implant restoring the eye's capability could be extended technically to be able to receive data beyond the ends of the visible spectrum of electromagnetic radiation. This could make it possible to create a visual prosthesis allowing us to see in the dark, as if using a night vision device. By using an artificial accommodation system, an optical zoom could even be integrated into the implant. The capacity to enlarge sections of a perceived image at one's discretion would presumably be quite attractive for many professions such as soldiers, birdwatchers, pilots, or surgeons. These ideas can be derived from existing technical possibilities, where they have been employed outside the human body, as in cameras, microscopes, and telescopes. Technological advance, which usually happens outside of the human body, can be redirected for technical intervention and internalization (Siep, 2006, 308).

Cognitive enhancement (Sarewitz and Karas, 2006) aims to extend the cognitive functions of the brain. If the brain is modeled as a machine that stores and processes data, these functions could be enhanced. On the one hand, the *storage function* of the human brain could be improved, e.g., by implementing a brain chip to save the information stored in the brain like a back-up device. By connecting a chip directly to the visual nerve, it could be possible to record and externally store all visual impressions in real time. In this way, all the visual impressions acquired in the course of a human life could be recalled at any time.

The *processing* of data in the brain could also be made subject to enhancement considerations. A chip implanted in the head and connected to an external network could be expanded as a new interface with the brain. It could be used, for example, to "upload" the content of books directly into the brain, or a device could be created that could load different language modules onto this chip and activate them as needed, making the tedious work of learning a foreign language superfluous. While such thoughts are pure speculation, they influence mindsets and motivate research activities.

The military especially exhibits great interest in enhancement and speculative possibilities. Relevant projects of DARPA, the advanced research arm of the American Department of Defense, are, for example, supposed to contribute, in the short term, to the following (following Coenen, 2008):

- Revolutionizing prosthetics by utilizing brain activity to control assistive technologies and developing limbs with full sensory and motor functionality.
- Developing systems with which computers would significantly enhance a soldier's performance, in particular under stress and while maintaining control of numerous devices.
- Enhancing the performance of soldiers affected by sleep deprivation.

Even further-reaching possibilities for technical human enhancement are discussed, such as controlling jet fighters solely by means of brain activity (projected for 2045).

(2) *Extension and alteration*

In addition to enhancing existing human functions, ideas are discussed for implementing completely new functions in human bodies which have no existing model in the natural human organism: "This means adding new features to brain functions (brain-to-brain interface, web access, etc.) through technological means unknown so far" (Jotterand, 2008b, 18). New sensors could, for example, be developed making use of Earth's magnetic field to improve orientation in problematic areas. The sonar and echolocation capability of bats could also serve as a model for implementing new capabilities:

The fundamentally novel quality of these thoughts about human enhancement is the fact that it no longer concerns an improvement of the body's own functions and performances but the imitation of performances by nonhuman beings and devices (Siep, 2006, 311).

Suitable neuroelectric interfaces could make it possible to connect still-to-be-invented motorized terminal devices to the nervous system providing completely new functions to humans. Such devices would be in addition to our arms and legs and should be controlled directly by the brain. Another aim is the development of exoskeletons steered by the brain, which make normal movement possible despite heavy loads, or bionic equipment with which soldiers can climb walls like a gecko or the movie figure Spiderman without any of the usual climbing equipment. This could provide attractive supplementary competence for training purposes and certain occupational groups: "Cognitive enhancements are envisioned by many as a future component of education [...] and by some as a future component of jobs" (Wolbring, 2008b, 28). It is however unclear how the human brain would process the information about novel organs and what effects this extension of our senses would have mentally if, for example, in addition to our natural limbs a third arm were to be controlled in the form of a robot by neuronal signals from the brain.

There is no limit to the fantasy of creating new ideas. As soon as the traditional shape of the human body is left behind, the door will be opened to the complete alteration of humans (Jotterand, 2008b). This development would go in the direction of the self-transformation or re-invention of humans, or the transformation of humans based on the ideas and power of a few. Even though such alteration frequently stands at the center of thoughts about enhancement because it provides by far the greatest challenges, this case will not be considered further in this chapter. The reason for this is simply that these considerations are completely speculative (Nordmann, 2007a).

(3) *Anti-aging and immortality*

Many attempts have been undertaken to extend the human life span. Some thinkers, like German philosopher Friedrich Nietzsche, perceived the limited life span of humans and the necessity to die as a scandal that should be overcome by science and technology. Since the beginning of the twentieth century, activities in this direction have increasingly met the requirements of modern scientific standards. The hopes to achieve longer life and reduce the negatively felt effects of aging have also spurred developments as diverse as the increase in sporting activity among the elderly, the creation of a market for new dietary supplements, the popularization of Botox medications as anti-wrinkle applications, or the upsurge in cosmetic surgery. In more recent times, new expectations for science and technology to prolong human life are related to the story of the NBIC converging technologies that in part played a role in Drexler's (1986) visions. These stories still exert a strong fascination (Sethe, 2007).

Ideas and expectations that aging might be slowed markedly, or even stopped entirely, play a central role in the discussion of human enhancement (e.g., Bainbridge, 2004). Such hopes are being nurtured by several developments in nanomedicine, which however, is supplemented by rather speculative assumptions. Among the most relevant expectations is a diagnostic procedure supposed to permit continuous detailed monitoring of a person's health status (Freitas, 1999). Procedures based on nanotechnology are being developed that enable carefully targeted drug delivery to affected parts of the body. These procedures alone would in many cases make substantially earlier and more efficient treatment of degradation processes possible and thereby contribute to a longer life for many people. These procedures would produce far fewer side effects than the classical therapies, which in comparison appear decidedly coarse. If aging is a degradation process at the cellular level – an understanding that is a matter of real controversy in medicine – then aging could be retarded by immediately discovering and repairing any manifestation of a degradation process.

Even further-reaching ideas are in circulation, which led to hype in the early years of the 21st century. The most famous idea was based on Drexler's concept of a molecular assembler (Drexler, 1986) and on the associated proposal to build nanorobots. These could be used, for example, to install a second immune system in the human body consisting of nanorobots. Such intelligent nanomachines could move through the bloodstream, serving as a technical immune system monitoring the human body in order to constantly maintain an optimal health status (Drexler, 1986). According to these visions, any degradation and every sign of physical decline should be recognized immediately at the atomic level and be stopped or repaired. In this manner, the machines might succeed in healing injuries or any other perturbations perfectly within a short period of time, ultimately slowing down and stopping aging.

Whether such visions can be realized, whether they are possible in principle, and how long a period of time is expected to be necessary for visible advances is highly uncertain. In 1967 Hermann Kahn predicted that in the year 2000 the average life span of humans in the US would lie between 100 and 150 years. In spite of many advances this has not happened. Yet some scientists consider it possible that the human life span could be extended from the current medically-accepted limit of 120 years – which was already mentioned in the Bible – to 250 years within the next few decades. Regardless of their speculative nature, it is important to pay attention to the effect that such expectations might have on public opinion, especially considering the anti-aging boom. Even if, behind the speculation, there were no possibility for such ideas to become reality, they could have real consequences for expectations, humans' self-image, and the agenda of science. Despite the speculative nature of such visions of immortality, ethical reflection began early in the debate. As an example, Moor and Weckert (2004) considered the question of what "quality of life" could mean in a society where humans could expect a life span of 500, 5000, or 50,000 years.

Another speculative vision in this context is that it might be possible in the future to download and store the contents of a brain's consciousness on a computer by means of a neuroelectric interface, and thus to transfer human consciousness to a machine. Since this machine can be repaired over and over again in cases of malfunctioning or technical aging, it would in a certain sense achieve a state of immortality. Another possibility would be to upload the consciousness from the computer device to new and other bodies. In this way, immortality could be combined with selecting a new body out of a range of options available. Again, these ideas are highly speculative but have some audience among transhumanists and in the mass media. The more recent wave of digitalization has fueled this type of expectation in certain circles (Chap. 8).

7.1.3   Enhancement in Human History

The dissatisfaction of humans with their own capabilities and properties is presumably as old as human history. Dissatisfaction with one's physical endowments, physical and mental capacity, dependence on external events such as illnesses, the inevitability of aging and ultimately death, or – and this is presumably particularly frequent – his or her appearance, are well-known and virtually daily examples of an individual's self-experience and represent furthermore the general self-experience of humans throughout history. The experience of one's own inadequacies when confronted by one's own or society's expectations, as well as one's limited capacities in dealing with challenges and the blows of fate – which is frequently perceived as inadequate – can be found in a wide spectrum of human self-experience. This self-experience extends from annoying forgetfulness concerning both harmless as well as important items of knowledge, to a collective experience of inadequate morality, such as in conflicts and wars, and of moral failure in view of the temptations of power or wealth, or simply for the sake of convenience.

Stories, fairy tales, and sagas process this experience, such as stories of the fountain of youth, or legends in which humans acquire superhuman powers. Spiderman and his colleagues of both sexes are modern expressions of such dreams. Cultural practices were developed to compensate for perceived deficits such as beauty flaws, for example, by means of cosmetics or clothing, and forgetfulness, by means of crib sheets (whether paper or electronic). Superhuman capacities – i.e., those clearly surpassing the average capacities of humans – have been and continue to be achieved by means of training, such as in sports or the ability to play a musical instrument. Even a cultural achievement such as our judicial system can be interpreted as compensation for humans' experience of deficits and as an aid in stabilizing civilization, regarding its fragile nature. Considering a human to be a "deficient being" (Gehlen, 1986) who builds their own civilization, including technical opportunities, to compensate for these deficiencies is a prominent anthropological interpretation of these experiences – although one to be criticized (Gutmann, 2004). A new manner of experiencing deficits is itself linked with technical progress. This is the experience of not being able to keep up with technical progress, of feeling inferior in view of the possibilities of technical systems, and of experiencing oneself to be antiquated even relative to one's own technical creations (Anders, 1964). This feeling spread to large parts of the population in some industrialized countries in the accelerated digital transformation (Chap. 8).

As a branch of the economy with substantial and increasing revenue, cosmetic surgery is today probably the most commonly used form of compensation for features perceived by an individual to be deficits. What does not please and does not correspond either to one's own expectations or to external ones is adapted by technical means. Doping in sports, which has led to repeated public scandals over several decades, represents the attempt to enhance (certain) humans. Such doping, however, is not only a cause of concern from the perspective of sports ethics (e.g., Hardman and Jones, 2010) but can lead to serious damage to the health of those involved and even to their premature death. Such doping continues to be developed despite its health impacts, social ostracism, and threats of relevant sanctions. The enhancement of certain individual abilities is also on the advance in everyday matters. Pharmaceuticals are one example, specifically those that are supposed to enhance the performance of individuals, such as in examinations (Sec. 7.4).

While the forms of compensating for (or overcoming) deficits in specified properties or capacities refer to *individuals* (athletic performance, personal beauty, good results in examinations), the *collective* enhancement of humans is certainly not a new topic. Even in the early modern period, there were breeding (!) ideas for enhancing humans, some referring to Plato's Republic (after Siep, 2006, 309). Frequently, complaints were expressed about deficits of humans from the perspective of morality and civilization, e.g., in the famous frame used by Thomas Hobbes' *Homo homini lupus*. The European enlightenment embodied its belief in progress as including a moral sense in approaches that attempted to employ *education* in a dedicated manner to enhance humans *as a whole*, i.e., ultimately enhance humans' and society's constitution in general. Beginning with the individual, above all through school and university education, a far-reaching higher development of human culture, civilization, and morality was to be stimulated and supported. For example, the critical theory of the Frankfurt school focused on the emancipatory function of education and influenced school curricula in many Western countries in the 1970s.

In totalitarian regimes, human enhancement was at the service of the respective ideology, rendering the term "enhancement" both problematic and ideological. In Nazi Germany and in the context of its biologically racist ideology, for example, enhancement included biological measures for strengthening the allegedly Aryan ideals. The ideal was represented by physical features (blond, blue-eyed, athletic) in connection with an unconditional subordination under the Nazi regime. "Human breeding" was its declared program with regard to the features that were accessible biologically, while with regard to social qualities the multiple possibilities of indoctrination and propaganda were utilized for what was understood to be "enhancement" in that totalitarian and racist sense. Stalin and Mao Zedong also employed propaganda and indoctrination in order to "enhance" individuals according to the orthodox communist ideology. In their sense, individuals should stop being individualists and become functioning parts of the collective without raising questions, putting "enhancement" into a more than problematic light.

The idea of human enhancement is on the one hand closely linked with individuals' experiences of their own individual and collective imperfection. On the other, we can see totalitarian misuse of the term "enhancement" for making people function only as parts of a totalitarian system. Therefore, the idea of enhancement stands in an at least ambivalent tradition. The negative experiences of enhancement fantasies and the measures employed by totalitarian regimes in the twentieth century to actualize them demonstrate that approaches to human enhancement must be scrutinized carefully in order to prevent totalitarian traits. Human enhancement in favor of ideological ideas of *future* society and *future* humankind can be used to suppress *contemporary* humans and to devalue or neglect *contemporary* desires and needs. Historical experiences force us to be careful about proposed enhancements and to ask deep questions about their motivations and possible side effects. The debate around human enhancement, however, has taken little notice of this history.

Indeed, the simple use of the word "enhancement" must be examined carefully. "Enhancement" is not a one-place predicate, because "enhancement" is not undertaken in itself but always only relative to certain criteria and contexts (Sec. 5.2, see also below). Enhancement relative to certain and ideological criteria (e.g., practicing Stalinist orthodoxy) can mean heavy deterioration relative to other criteria, for example, with regard to the ideal of free and emancipated citizens. Caution is thus due in order not to fall into the obvious rhetorical trap of everyday language, namely of considering an "enhancement" to be positive in itself. Instead, we must inquire as to the criteria according to which modifications are regarded as an "enhancement" and why this was or should be done.

7.2 The Semantics of Human Enhancement

We have already discussed *animal enhancement* earlier in this book (Chap. 5). In that chapter, a semantic clarification of the term enhancement was provided (Sec. 5.2) which will also be used for discussing *human enhancement*. However, further semantic and philosophical clarification is needed before applying it to humans. First, as already indicated in the previous section through specific examples, we have to take care of both the difference(s) and the boundary(s) between medical healing and enhancement, while also considering possible transitions between them (Sec. 7.2.1). Second, the issue of doping occurs repeatedly in the debate on human enhancement. This requires reflection on the differences between healing, doping, and enhancement (Sec. 7.2.2). Finally, it is necessary to specify the distinctively *technical* aspect of human enhancement (Sec. 7.2.3).

7.2.1 Enhancement *versus* Healing

The conceptual and terminological field between healing and enhancing, improving performance and doping, and modifying human nature is the object of philosophical debates (e.g., Jotterand, 2008b; Hornberg-Schwetzel, 2008). Clarifications must be provided in advance in order to identify the need for ethical reflection, since, depending on the concepts, different normative contexts and possible uncertainties might be involved. For example, healing is covered by medical ethics while enhancement exceeds established medical ethos (e.g., Beauchamp and Childress, 2013). First, the main results of the early discussion of the semantics of enhancement will be recalled (cp. Sec. 5.2 for details).

The term "enhancement" sounds fundamentally positive. However, its semantics is much richer and more complex than is often assumed, because enhancement can only be determined relative to certain criteria. Enhancement represents an activity or action by which an object is changed in a particular direction: there are *actors* (the subjects of enhancement), who enhance *something* (the object of enhancement) according to *criteria*. This activity necessarily includes the semantic dimensions of a *starting point* for enhancement, the *criteria* of what is regarded as enhanced, and a *measure* of enhancement indicating the difference between the starting point and the status reached by the enhancement activity. Many ethical challenges of human enhancement emerge out of this constellation; for example, because an enhancement relative to one criterion can be a deterioration relative to another, as is well known from technical developments. Furthermore, the difference between the goals of an enhancement measure determined *ex ante* and the actual consequences of the measure *ex post* must be noted. Possible failures and unintended consequences have to be taken into account. Therefore, complex assessments of relative value and risk are necessary before a judgment and decision on an enhancement measure can be made. This assessment must involve the *goals* of the enhancement, the *criteria* oriented to these goals, the measure for the intended extent of the enhancement, the *deteriorations* that may be linked to a specific enhancement from a different perspective, and finally the possible manifestation of *unintended consequences* of the enhancement measures (Sec. 5.2).

Most of these challenges are well known in the field of medical healing. But while healing is covered by the normative framework of medical ethics developed over human history, human enhancement raises new questions and causes normative uncertainties. Therefore, we have to clarify the relation between healing and enhancement, which is not that easy: "The accent in the discussion of bioethics today is usually on the question of whether it is at all possible to draw a line separating therapy and enhancement" (Siep, 2006, 306). Yet both terminologically and in everyday understanding, human enhancement is categorically different from healing an illness or compensating for the consequences of an accident. Healing is oriented toward the regulative idea of a healthy individual. Healing reaches its goal as soon as the patient becomes healthy even if this may be difficult to determine in detail. Enhancement, however, does not contain any stop criterion within itself.

This does not mean, however, that empirically – i.e., on the basis of medical or biological data – healing can be fundamentally and unequivocally distinguished from enhancement in each case (e.g., King and Hyde, 2012). Judgment about this distinction is based not only on data but also on normative criteria of what is regarded as healthy and when a status reached is "healthy enough." The concepts of "health" and "illness" are subject to positive and negative valuations of conditions depending on not only medical or biological data but also on sociocultural contexts and normative criteria (Engelhardt, 1976; Engelhardt, 1982; Margolis, 1981; Nordenfelt, 1993). They are thus defined by making reference to both measurable data and value judgments; they combine both descriptive as well as normative components. Consequently, the attributes that a healthy human is supposed to possess cannot be determined by natural and medical science alone, although scientific data are relevant. It is therefore impossible to draw a clear distinction between healing and enhancement at the level of biological or medical data. Instead, responses to this challenge depend to a certain respect on history, ethics, and culture (Hornberg-Schwetzel, 2008).

Often, transitions between healing and enhancement are possible. Healing – i.e., the restoration of certain bodily functions – may contain an element of enhancement. An unintended (and often positive) side effect of healing can be an enhancement in a different respect. The replacement of limbs by artificial implants can lead not only to a restoration of lost bodily functions but also to enhancements, such as the higher mechanical stability of artificial limbs or a longer operational lifetime. Pistorius, the athlete amputated below the knee, faced exclusion from international competition for this reason (Wolbring, 2008a). In this way, i.e., as the unintended consequence of acts intended to heal, enhancements can enter routine medical practice and could potentially be normalized in an insidious manner. Another example of the context-dependency of this assignment relates to the drug Ritalin. While this is used as a drug to enable children with ADHD (attention deficit hyperactivity disorder) to better concentrate, in the realm of overcoming a deficit, it is also used as a drug for weak enhancements when applied to persons without ADHD, e.g., to students in an examination.[1] The same measure may be classified as healing or enhancement, depending on the context of application.

Contextually, however, the distinction between healing and enhancement usually is quite clear because healing addresses certain goals and should stop as soon as the goal is reached. The ophthalmologist who gives a patient an eye test has an understanding of what a healthy human eye is capable of performing. They will only suggest technical compensation (e.g., glasses) if the results of the test deviate from the standard expectations and reaches a certain magnitude. The goal of this measure is to reach the statistically determined normal performance of human eyes. Healing ends when this state has been reached. It is impossible to imagine traditional medical conduct without the regulative idea of a presumed normal state regarded as healthy for recognizing deviations and diagnosing the necessity for interventions. Even though the boundary between healing and enhancement at the object level of medical data is controversial and may be unclear, clarification can be found in the rules and criteria inscribed in medical practices in a specific culture at a certain point in time.

The discourse of healing is different from that of enhancing, and the normative frameworks of both discourses differ distinctly (see below). In the realm of healing, the medical oath is relevant and supported in reflection and analysis by medical ethics. Therapy is subject to a sense of obligation, advisability, or expectation – an *Ought* – inasmuch as its purpose is the (re)creation of biological functioning based on an understanding of being healthy that is adapted to a specific culture. Talking about human enhancement, however, has no regular place in established medical ethics. In the case of enhancement, there is no *should* (Siep, 2006) but only or at best a *desire* (Hornberg-Schwetzel, 2008). To put it differently: the boundary between healing and enhancement cannot be determined by medical or biological measures or data. Instead, this boundary is constituted by different normative frameworks separating different

---

[1]In a medical respect, there is only low evidence for real enhancements of healthy humans, if any (Sauter and Gerlinger, 2011).

discourses rather than constituting a boundary between different physical properties of humans or of interventions into humans.

7.2.2 The Semantic Field of Healing, Weak and Strong Enhancement

Clarification of the meaning of human enhancement will be built by first looking at possible *points of departure* for enhancement processes and thoughts (Sec. 5.2). Imagine a situation in which measures to be applied to an individual person are discussed, whether they belong to the area of enhancement or not. The point of departure, if looking at enhancement, could possibly be one of the following options:

(1) The physical or mental endowment of this specific human *individual.*

(2) The standard of an *average* healthy person, as measured for instance according to statistical surveys of human performance potential.

(3) The achievement potential of humans under optimal conditions, that is to say, the achievement potential at the *upper end* of the statistical distribution.

Let us take again the example of a person visiting an ophthalmologist. In the first case, even a pair of glasses would be an enhancement for that individual, if they, for example, are short-sighted. If we apply the second option, this pair of glasses would not be an enhancement because this measure would not enhance the patient's capability beyond the standard of a healthy human according to customs and criteria valid in the respective situation. This would also apply to the third option. In this case, one would ultimately only speak of enhancement if the abilities achieved exceeded what is typical at the upper range of the performance spectrum of all humankind.

Regarding the reflections given above on the different normative frameworks ruling healing and enhancement (Sec. 7.2.1) it is quite clear that option 1 describes a typical process of healing. Therefore, this will not be subsumed under enhancement simply because it is governed by established medical practices according to cultural agreement about what is regarded as standards of health and customary human capability. In cases of normative uncertainty in this field, established medical ethics are prepared to help regain orientation.[2]

Option 3 clearly exceeds the normative framework of medical ethics. Enhancing humans in this sense means improving some of their capabilities beyond the capability of humans under optimal conditions. Corresponding directions of enhancement could be – according to the ophthalmologist example above – to extend the perceptibility of electromagnetic radiation with respect to wavelength, to enlarge the storage capacity of the brain, or abolish the risk of forgetting knowledge and memories. Generally, those modifications would be regarded as human enhancements that in some manner make humans more powerful than is expectable under optimal conditions, thereby empowering humans to act beyond the upper end of the statistical distribution of human performance.

Option 2, however, lies in between, which can be observed by looking at the field of sports. Training methods used in sports, coaching techniques, etc. do not fall under strong enhancement measures in the sense mentioned above. These measures aim at improving the capabilities of individuals toward the upper end of the statistical distribution of the capabilities of humans, e.g., with regard to sprinting in athletics, but without exceeding the upper end of this distribution. Sometimes, perhaps, the upper end might be moved or pushed by incremental steps, which, however, does not create normative uncertainty because of their incremental nature and embeddedness in the normative framework of the ethics of sports. Normally, cosmetic surgery also does not serve to enhance humans in the sense defined above, at least as long as its aim is to achieve accepted ideals of beauty, i.e., those which are not out of the ordinary within cultural development and history. Doping in sports or for any other contest is neither healing nor strong enhancement according to the clarifications given above but rather something in between. To make the linguistic usage more precise and at the same time draw attention to each of the different normative frameworks that regulate these options, I will propose the following terminology

---

[2]Particular challenges occur for analyzing and assessing the relation between disability and health (e.g., Wolbring, 2006).

to distinguish between weak and strong enhancement, based on the distinctions made above, thereby making earlier distinctions (Grunwald, 2012) more precise:

Healing: The elimination of an individual's deficits, such as healing diseases and restoring lost capabilities relative to the accepted standards of an average healthy human being, is considered to be *healing*, in accordance with the tradition of medicine. Determinations of what is regarded as an accepted standard and as average health are not determined by medical or biological properties alone but will also depend on cultural and social issues (see above). Normative uncertainties can be tackled by medical ethics. Measures in this field will explicitly be excluded from further consideration of enhancement, except for illustrative purposes.

Weak enhancement: An increase in a healthy individual's performance, beyond both their individual capabilities and the accepted standards of an average healthy human being but without going beyond what is still regarded as conceivably normal, i.e., within the spectrum of usual human performance under optimal conditions, is demarcated as *weak enhancement.* This may apply to doping in sports but also to other changes, such as cosmetic surgery to improve personal appearance beyond the ideals of the average.[3] The postulate of fairness and risk ethics provide normative frameworks to gain orientation in case of uncertainties.

Strong enhancement: An increase in performance beyond abilities that are regarded to be normally achievable by healthy, capable humans who are ready to perform under optimal conditions will be denoted as *strong enhancement.* In other words, and recalling the clarifications above: applying measures that lead beyond the upper limit of the statistical distribution of a specific human capability or that empower humans with completely new capabilities are subsumed under *strong enhancements.* The latter is

---

[3]In the field of cosmetic surgery, the boundary between healing and doping will be much more difficult to determine compared to other medical practices, because it is highly controversial what "average" should mean here. However, this distinction is made in contextual medical practice, regulation, and the reimbursement policies by insurers. Decision-making is oriented by certain criteria determined within the insurance system, indicating that there is some implicit idea about the boundary between healing (will be reimbursed) and enhancement (will not be reimbursed).

sometimes given its own characterization as *alteration* of the human composition, for example, by implementing new organs or bodily functions (Jotterand, 2008b). It will be subsumed under strong enhancement throughout this book because it leads beyond the normative frameworks of medical ethics and the ethics of sports and thus contributes to similar normative uncertainty. In both cases, no established normative framework is available to create orientation.

The distinction between weak enhancement as performance-improving in the sense of option 2 and strong enhancement in the sense of option 3 is essential in the interests of a clear use of language. This holds in particular with regard to the respective normative framework governing these areas, and thus with regard to the possible normative uncertainties and the necessities for ethical reflection. The ethical problematic of weak enhancement, e.g., doping and related normative uncertainties may be characterized above all by two normative demands on human action or, respectively, by the problems arising from the fact that these demands are often insufficiently met: the demand for fairness and the demand that one acts responsibly toward one's own body (e.g., King and Hyde, 2012):

- *Fairness* is the essential demand in the field of sports that is violated by doping. Individuals who dope gain unpermitted advantages over those who follow the rules. This situation is well known from many areas of sport and applies to many other fields of human activity, as will be shown below.

- Doping often poses *risks* to the human body. Short-term threats to health may arise. From a legal point of view, sports doping is mainly the abuse or misuse of drugs, often far beyond their approved use. Completely unknown long-term risks can also occur. Cosmetic surgery as a kind of weak enhancement to improve appearance repeatedly causes health risks. Both types of risk raise questions about the responsibility of individuals for their own health and life, weighed against short-term interests in becoming better in e.g., sporting contests, which is sometimes related to economic interests.

Taking the distinction between weak and strong enhancement seriously results in clear implications. As already mentioned, sports doping as we know it today does not fall under our definition of strong enhancement. Everyday doping, such as taking stimulating pharmaceuticals such as Ritalin (Farah *et al.*, 2004), e.g., prior to examinations, is also not to be subsumed under strong enhancement, unless it leads the individual to exceed the customary abilities of test candidates to an inordinate degree. Even steps in cosmetic surgery would therefore not be considered as strong enhancement; these fall partly under healing and partly under weak enhancement in the sense of "beauty doping" (see footnote above). The goal of the latter, analogous to the goal in sports doping, is for individuals to gain advantages, e.g., by improving their chances in beauty contests, or just to feel better.

For weak enhancement, including beyond sports (see below), existing normative frameworks can be applied, such as the ethics of sports, or postulates of fairness in competition in the labor market. For example, the ethics of sports coupled with considerations of prevention, education, and sanctions is appropriate to gain orientation, e.g., for gene doping (Gerlinger *et al.*, 2008). This does not apply to strong human enhancement, i.e., in terms of the production of "superhuman" or "transhuman" capabilities and features. In cases of weak enhancement, topics that dominate the enhancement debate, such as human self-image and the future of human nature, are not touched upon. The normative uncertainties involved in both weak and strong enhancement thus *differ by category*. For this reason, it makes sense to draw a strict line between weak and strong enhancement in ethical reflection (Sec. 7.3). There is hardly any overlap between considerations of how we should handle the consumption of drugs to incrementally improve brain performance and questions about the direction a society would take if the ability to discriminate between humans and technology was (in the future) largely removed.

The following example illustrates the dependence of judgment on whether a certain measure is classified as weak or strong enhancement. If the drug Viagra is consumed in case of erectile dysfunction, then this is considered in the context of eliminating deficits, as measured by the customary attributes of humans, and thus belongs in the realm of healing. If it is consumed by a healthy person in order to increase the duration of erection, then this is an improvement of performance above the normal level. If the increase in duration of erection remains in a customary range but nonetheless constitutes an improvement for the individual, then it has to be classified as a type of doping and would be considered an enhancement in cases of extraordinary prolongations of the erection. This illustrates that the attribution of the terms healing, weak and strong enhancement depends on interpretations of the respective situation, and in particular on what "customary" or "normal" is taken to mean. The respective attributions simultaneously relate to different normative frameworks governing the respective field.

The boundaries between these three categories may be a matter of controversy. Even though these boundaries may not be naturalized at either the object or data level, they exert a decisive influence on the ethical debate. Each is involved in a different normative context: in the case of healing, medical ethics; in the case of weak enhancement, the fairness imperative and considerations of risk; strong enhancement enters a largely normatively uncharted area (Sec. 7.3).

7.2.3 Human Enhancement by Technology

Throughout history, humans have always looked for improvements with respect to technological means, which usually affected the environment of humans but not their own bodies. Humans have enhanced themselves not by technology but by education with respect to culture, e.g., in the European Enlightenment. This idea of internal enhancement by education however, has also been misused for ideological reasons, by replacing education with propaganda or pressure (Sec. 7.1.3). Enhancing humans by technology is an idea from recent decades. In order to understand its meaning and implications, technical enhancement in engineering serves as a model.

Engineers know from their daily practice about technical enhancements. Every technology can be described by certain parameters, which include its performance characteristics. Enhancement would mean improving one or more of the already attained performance characteristics according to the accepted standards of technology, such as motor performance, efficiency, environmental compatibility, life span, or price. The *starting point* of the enhancement consists of a description of the respective state of the art of the technology to be enhanced. The criteria and direction of the enhancement result, e.g., from considerations of the company's management in order to better fulfill customers' desires or to decrease production costs. As a rule, the magnitude of the enhancement can be determined directly by comparing, usually quantitatively, the values of the corresponding parameters before and after the enhancement. In evaluating the result, possible deteriorations of other parameters must be included. This approach to enhancement follows the model taken from engineering. For its realization, techno-morph modeling of humans is necessary (Chap. 9).

Engineers know that technical enhancements frequently represent deterioration from *other* perspectives. Specialization based on one specific criterion of performance often leads to deterioration according to other criteria. For example, more safety often means higher costs. In human enhancement, this ambivalence concerns first the increased dependence on functioning technology, in this case on functioning enhancement technology. Technical enhancements which modify certain human performance characteristics that can be described in a techno-morph manner, lead to a specialization that is diametrically opposed to the multi-functionality of natural human limbs and sensory organs. An athlete, for example, who has artificial legs implanted to improve their performance in track and field events, will very probably have subsequent problems swimming that they would not have had without the technical enhancement. The fundamental ambivalence of technology (Grunwald, 2019a) is thus transferred into the center of the human realm by human enhancement.

A second issue migrating from technological advance which is external to humans into the human body and mind is the idea of an infinite advance by continuous improvements in engineering. For an illustrative example, cochlear and retinal implants as technical compensations for deficits, such as capacities lost as a result of illness or accident, will be considered. There has been great progress over recent years in restoring parts of the ability to see and to hear. Further progress is expected. The ultimate goal – fully corresponding to the medical oath and medical ethics – is to employ technical implants to re-establish the patient's full level of performance by reproducing the natural functioning of the body. Ideally, the implants should be so small that they would be invisible to others.[4] In this case, a criterion for reaching the goal of an identical result could be, for example, that an ophthalmologist has the implanted patient take an eye test, objectively examines the visual performance, and compares this result with the standard set by an average natural eye. Developing such technical implants is covered by the familiar normative framework of medical ethics.

Now, let us assume that a university institute or a company succeeds in engineering, for instance, a sensory organ such as an eye with equally good results as compared to an average human one. An artificial eye of this kind would be given, as is customary in engineering, a version number by its manufacturer. This would be "eye 1.0." Version 1.0 will not be the final one, however, because engineers and physicians will be thinking of the next version as soon as version 1.0 has been developed and tested. Continuous improvement is a technological imperative in modern engineering. The subsequent improvements might take a slightly or entirely different direction, e.g., a reduction in costs or a prolongation in the service interval for eye version 1.0. Although the human sensory capabilities of version 2.0 consequently do not have to be improved (e.g., by providing night-vision ability or zoom options), continuous improvement nonetheless forms part of the spectrum of the technological imperative, as applied to eye 1.0. Human technical enhancement is thus revealed as a step, which will be followed by the next step, and so on. Therefore, the transition from *restorative* to *improving* interventions is a gradual one *from an engineering perspective*. The technological imperative leads *of necessity* from healing to improvement, either for doping or enhancement purposes, if normative arguments do not restrict it. The necessity is deeply inscribed in the engineering culture, tradition, and thinking that has developed over the last two centuries. Here, the idea of an unlimited technological advance meets human beings. Modeling humans in a techno-morph manner for applying technical enhancement

---

[4]This does not mean that the technical processes employed for this purpose have to reproduce the natural ones; the opposite will usually be the case. For example, a retinal implant does not have to use the same mechanism to detect incident light as the natural retina. Yet the outcome must be an identical result, in the sense that the data received by the sensors and passed on to the brain need to produce the same impressions that the natural organ did.

measures, therefore transfers the idea of unlimited technological advance to the idea of unlimited enhancements of humans by technology.

Human enhancement then, has no intrinsic limits or measures but opens up an infinite space of the possible, just as technical progress does in general. Enhancement does not aim to achieve an "optimal" final condition. Being open with regard to a final condition implies that there is no inherent limit to the enhancement: the result of each enhancement can serve as the starting point for new enhancements, possibly based on different criteria. The process of enhancement is not complete and will never be complete. The process of human enhancement does not stop once a given status has been reached, i.e., in the sense that a target has been reached. It serves on the contrary as the starting point for the next enhancement, and so on. Technical enhancement of humans, similarly to technological advance, would be an infinite process if there were no external societal measures for "containment" or for determining limitations. Limits to enhancement can, for example, consist in ethical arguments (see below), or in acceptance or other social issues.

7.3   Ethics of Human Enhancement

Human enhancements, strong or weak, leave behind the normative framework of medical ethics and the area of healing and restoration. New fields of human intervention open up for which normative conventions have not yet been established, particularly for strong enhancement. In this section, normative uncertainties will first be made an issue (Sec. 7.3.1), followed by a discussion of the major fields of ethical inquiry which respond to them (Sec. 7.3.2). Finally, the ethical debate so far will be summarized (Sec. 7.3.3).

7.3.1   Normative Uncertainties and Ethical Questions

Strong and weak enhancement of humans constitute steps into a new sphere, not only in a physiological but also in a normative sense. The fact that normative customs and moral rules did not yet exist was quickly identified as a clear challenge to ethics, at least for strong enhancement (Farah *et al.*, 2004; Khushf, 2007; Ach and Siep, 2006).

To understand these challenges more precisely, the first task is to identify and characterize the normative uncertainties that have arisen. Expressed abstractly, this is simple. If the human body and mind could be formed technically, then the question arises how far people *may*, *should*, or *want* to go in (re)forming the human body and mind for enhancement purposes. Controversies surrounding the answers to this abstract question emerged at the very beginning of the enhancement debate, directly after publication of the NSF study on converging technologies (Roco and Bainbridge, 2002). While transhumanist positions (e.g., Savulescu and Bostrom, 2009) see not only a moral acceptability but even an *obligation* to achieve enhancement, other authors warn against this, frequently giving ethical reasons for their position (e.g., Siep, 2006). To structure the field of normative uncertainty questions, the following classification has been suggested for cognitive enhancement (Sarewitz and Karas, 2006, i):

- Laissez-faire: emphasizes the freedom of individuals to seek and employ enhancement technologies based on their own judgment.
- Managed technological optimism: believes that while these technologies promise great benefits, such benefits cannot emerge without government playing an active role.
- Managed technological skepticism: believes that quality of life arises more out of society's institutions than its technologies.
- Human essentialism: starts from the notion of the human essence (whether God-given or evolutionary in origin) that should not be modified.

Although this classification is generic and does not give justice to the full argumentations of the respective protagonists, it nonetheless provides a clear impression of the range of possible answers and the normative uncertainties of human enhancement. In the other field of enhancement, i.e., the fight against aging and the necessity of death (Sec. 7.1.2), it appears more difficult to take a position, since the extension of humans' life expectancy has been an object of the established medical ethos since its beginning. New challenges with respect to ethics will in particular emerge in a consequentialist manner when considering a possible world where humans live much longer than today (Moor and Weckert, 2004).

Against the background of fundamental issues in ethics, especially as to human autonomy, distributive justice, and dealing with the uncertainty of knowledge, the abstract question of what we should, want to, or may do in this field can be broken down into the following more concrete questions. These questions have developed into something like a canon of ethical questions directed at human enhancement (see, e.g., Williams and Frankel, 2006; Jömann and Ach, 2006; Siep, 2006; Jotterand, 2008b):

- What are the *criteria* according to which the enhancement will or is supposed to take place? Are all of the criteria ethically justified? How are decisions about these criteria made, and who decides? Is it possible to indicate an ethically grounded ranking of the priorities?
- Are enhancements in one sense linked with deterioration in another? How could the outcomes be weighed against one another? How can a measure of enhancements be determined?
- What risks to the individuals affected have to be taken into account, and what measures can be taken prophylactically to prevent the misuse of enhancement technology? Is informed consent sufficient, or can we imagine situations in which people seeking enhancement must be protected against themselves, as is the case in the field of doping in sports?
- Will the rights of those affected be touched on without them having been asked for permission or without this being possible?
- What are the consequences of human technical enhancement from the perspective of *just distribution* (Walzer, 1983)? What are the prospects for a deeper division of society into technically enhanced superhumans and unenhanced people who thus are in a position of inferiority?
- Will any enhancement options that are accepted devalue currently accepted and practiced forms of life such as that of the disabled? Or even make them impossible (Campbell/ Stramondo, 2017)?
- Could an obligation emerge to apply enhancement measures to disabled people in order to remove their disabilities and to establish more equity (Campbell and Stramondo, 2017)?

- What do we think of the heightened performance of individuals that has already become part of our society: cosmetic surgery, sports doping, and performance-enhancing pharmaceuticals? Do these fields of existing practices deliver experiences, role models, and paradigms also for other fields of weak enhancement (doping), perhaps even for strong enhancement?

- Can a spiral be triggered that creates a compulsion for individuals to seek more and more enhancement in order, for example, to remain competitive in the labor market? Could this lead to an "enhancement society" (Sec. 7.4), or have we already entered this?

- Will human enhancement push open an infinite space of further and further enhancements, as is suggested by the technological imperative (Sec. 7.2.3), or are there *limits* to enhancement? How can limits be explained, and how resilient are they argumentatively? On what grounds does their justification depend, and what are their premises?

- Should public funds for research be made available to scientifically develop enhancement proposals and to put them into practice? What role will economic arguments play here (this takes up a large space in Williams and Frankel, 2006)?

- How does human enhancement, in particular in its strong form, affect our images of human nature and of ourselves as humans? Could it lead to a technicalization of humans, and what would this mean?

- What can ethics do in order to help gain well-reflected orientation to respond to all these questions? How should ethics be performed in the areas of weak or strong enhancement?

Many of these questions touch upon fields of research policy and various scientific disciplines such as political science, regulatory issues, funding policies, anthropology, and psychology. However, all of them include normative uncertainties, implying that ethical reflection will have to play a role in finding appropriate responses.

These questions label certain points in a sometimes confusing debate that has been emerging since the beginning of the 21st century and is still developing. In the following section, we have to structure this material and to reconstruct major patterns of ethical argumentation that are already in use in order to gain an overview of the confusing state of the debate.

7.3.2   Ethical Argumentation Patterns

Moral challenges related to human enhancement were taken up by philosophical ethics quickly after the NSF report on converging technologies (Roco/Bainbridge 2002) was published (e.g., Khushf, 2007; Farah *et al.*, 2004; Siep, 2006; Jotterand, 2008b), based on previous reflections in the field of genetic enhancement of humans (e.g., Habermas, 2005, cp. Sec. 6.3). Various ethical patterns of argumentation were employed, but reflections from anthropology and the philosophy of technology also enriched the debates. The debate was, and still is, challenged by large uncertainties and the partially speculative expectations regarding technical possibilities and social consequences (e.g., Nordmann, 2007a). As a reaction, many of the reflections can be understood as thought experiments in methodological terms, assuming the appearance of empirical manifestations and consequences of enhancement and then examining these from an ethical perspective. The reflection on how a world would look if humans could have a life span of 500 or 5,000 years (Moor and Weckert, 2004) is a nice illustration of this "what if" approach. The concerns mostly mentioned in this "what if" manner are: loss of creativity, increasing concerns about security, the world's population increasing beyond all measure and consuming more and more resources, or even the question of how reproductive behavior will or has to change.

In this way, the argumentation patterns to be discussed below also mostly follow this "what if" structure, with "what" usually remaining rather speculative. Therefore, the ethical considerations do not end in practical orientation relevant to action or policy such as in the field of applied ethics. They are rather a tentative preparation of a new field of ethical reflection in the sense of explorative philosophy (Chap. 9).

Assuming that more and more forms of weak and strong human enhancement will become available in the future due to the scientific and technological advance, ethical analysis can be conducted by considering different issues and by applying various argumentation patterns. The patterns of argumentation proposed so far are incommensurable to a large extent. Anthropological thoughts about humans' naturalness or about the human–technology relationship cannot simply be balanced by ethical considerations of possible risk to human health or by statements about human rights.[5] The various patterns of argumentation are, therefore, discussed critically in the remainder of this section, without claiming to be systematic.

(1) *Human autonomy*

Human enhancement as considered above in this chapter addresses responsible adults who have provided their informed consent. Yet compulsion could develop out of the presumed autonomy of those affected in the course of the spread of enhancement technologies: "If neurocognitive enhancement becomes widespread, there will inevitably be situations in which people are pressured to enhance their cognitive capabilities" (Farah *et al.*, 2004, 423). The reason would lie in the fact that refusal of enhancement might be tied to disadvantages with regard to participation in social processes, such as in working life. Certain enhancements could become standard in certain occupational groups, just as today the possession of a driver's license is a prerequisite that is taken for granted among those seeking employment in many occupations. Someone who disregards the standard will have to reckon with disadvantages:

> With the advent of widespread neurocognitive enhancement, employers and educators will also face new challenges in the management and evaluation of people who might be unenhanced or enhanced (for example, decisions to recommend enhancement, to prefer natural over enhanced performance or vice versa, and to request disclosure of enhancement) (Farah *et al.*, 2004, 422).

Even if the autonomy of affected persons is maintained in this case in a legal sense, it is counteracted *de facto* by a compulsion on individuals to conform. A refusal to conform would have serious consequences. The creation of enhanced life as a new standard could devalue established forms or even ruin their survival. For

---

[5]In this chapter, I focus on enhancement measures applied to adult persons able to declare their willingness to be enhanced on the basis of an informed consent. The issue of enhancement by interventions into the human germline or into embryos is discussed in Chapter 6.

example, the way of life of certain disabled people could lose its social acceptance if it were possible to compensate technically for the disability (Wolbring, 2006; Campbell and Stramondo, 2017). This fear has already been discussed with reference to prenatal diagnostics in the context of permission for late abortions, possible genetic interventions to avoid disabilities, and the discussion of eugenics. In the context of human enhancement, this fear would pose itself in an intensified form.

(2) *Distributive justice*

The enhancement of an individual's human capacities would probably be tied to a high expenditure of resources, knowledge, and capital, which would strongly restrict the circle of beneficiaries. The ethical demand for an equal distribution of accessibility therefore will be challenged, or even appears naïve: "It is likely that neurocognitive enhancement, like most other things, will not be fairly distributed" (Farah *et al.*, 2004, 423). This raises the question of who could afford to undergo enhancement and what consequences the associated unequal distribution of opportunities for enhancement would have for those who cannot afford it:

> The ability divide will develop between the poor and the rich within every country. It will be bigger between low and high income countries than it will be within any given country. Not everybody will be able to afford enhancement of their body, and no society will be able to enhance everyone, even if they wished it (Wolbring, 2006, 31).

The range in the distribution of individual abilities that has always been present in society would be greater, as would the difference between the high performers and ordinary people. Enhanced individuals might even have the idea of securing their advantage or increasing it further by limiting the access of others to enhancement. The result might be a societal fault line between the enhanced "superhumans" and the normal humans. This concern has been regarded as a strong argument against human enhancement: "Securing such an evolutionary advantage over those who are previous conspecifics definitely represents a harm" (Siep, 2006, 318). The ethical reason behind this position is that this development would strongly violate the rights of the non-enhanced, i.e., those without access to enhancement technology. This development could also lead to new forms of eugenics: "The individually desired enhancement of human features and functions [...] is now conceivable, a form of 'liberal eugenics'" (Siep, 2006, 309). In addition, researchers who are more positive about human enhancement are concerned about the issue of distributive justice (Merkel *et al.*, 2007).

Shifting the standards of what is regarded as healthy and normal through widespread enhancement could shake the existing system of providing healthcare to its foundation. "Above all, this would eliminate the scale for measuring health and normality, which until now has been used to measure claims for help and compensation" (Siep, 2006, 320). The legitimization of medical intervention and/or the determination of reimbursement for medical services by a health insurance provider is currently oriented on healing to a large degree. This established type of orientation would disappear with widespread enhancement. Great effort would be needed to reorient the healthcare system on other normative ideas whilst observing standards of distributive justice about the role of medicine and health insurance systems.

(3) *Risks and unintended side effects*

Human enhancement measures do not necessarily succeed; they might fail and lead in specific cases to health problems or even to death. Examples are well known from the fields of doping in sports (Gerlinger *et al.*, 2008) and cosmetic surgery. This issue raises questions of accountability and liability, as they are known from the field of medicine. Yet even in the case of success, undesired effects might appear, such as dependence on the enhancement technology that could lead to helplessness in the case of failure, or psychic changes. It is difficult to identify *specific* concerns about enhancement compared to other fields of human activity in this respect: "And if we are not the same person on Ritalin as off, neither are we the same person after a glass of wine as before, or on vacation as before an exam" (Farah *et al.*, 2004, 424). However, established practices in medicine and cosmetic surgery can serve as models to deal with this type of enhancement risk. At minimum, affected persons must be informed as fully as possible before implementation of the measure. Another orientation is that the implementation of the planned enhancements should be executed in a manner to make these as reversible as possible.

(4) *Obligation to enhancement*

The primary issue discussed in the previous sections was whether human technical enhancement is *permitted* morally. The ethical question can also be posed for a possible obligation to enhancement, namely regarding *ought to* or *should*: are we obliged to pursue human enhancement if it became possible? Might enhancement perhaps even be ethically imperative because, for example, it would alleviate human suffering, because it might help solve great problems, and because it would make it possible to achieve a substantially greater quality of life? Is human enhancement perhaps imperative for reasons very similar to those for providing medical aid when someone is ill, e.g., for empowering disabled persons (Campbell and Stramondo, 2017)? Should we pursue human enhancement in order to secure the survival of human civilization in view of dramatically changing environmental conditions on planet Earth?

The main argument for the negative answer to the "should" of human enhancement (Siep, 2005; Siep, 2006) has been based on uncertainty issues: high uncertainty whether the goals associated with human enhancement will be reached as well as about the unintended consequences. According to this argument, a "should" cannot be justified. For the moment, this line of argument seems convincing. In particular, no one has indicated clear and reasonable purposes for which humans should be enhanced. Considering the nature of uncertainty, however, this can only be a *preliminary* answer. For example, specific requirements of certain professions could provide a normative background that enhancements should be performed in narrowly delineated areas. Yet, even if it were the case "that we want our firefighters stronger, aircraft pilots with clearer sight, and our researchers smarter, yet we want our athletes to play fair" (Williams and Frankel, 2006, 10), then criteria employed for decisions on enhancement could be derived as to who should and who should not be enhanced and for what purposes.

(5) *Moral status of cyborgs*

The debate on human enhancement by technological intervention into human body and mind is accompanied by questions about an increasing technicalization of humans. More far-reaching concerns about the emergence of a new species have also occurred: beings made from both human material and technical elements, being neither a human nor a technical object but cyborgs without a clear moral status. However, for the time being and for the foreseeable future, this presumed ethical problem is a pseudo-problem. A beautiful example of this is the fact that Ken Warwick, a known advocate of human enhancement through wearing neuro-prostheses, regards himself to be the first cyborg. He was also classified as a cyborg by others: "He is the world's first cyborg, part human and part robot. In 2002, he breached the barrier between man and machine" (Jotterand, 2008b, 23). This diagnosis, made in the cited article, is supposed to make it obvious that human enhancement will not only lead to the emergence of composite beings but is already a social reality in the form of at least one existing cyborg. This observation was supposed to demonstrate the urgency of ethical reflection, namely that it was high time for us to concern ourselves with the moral status of composite beings.

This example however, proves just the opposite and disavows all the excitement. The main reason is: Ken Warwick is not kept in a zoo or a high-security tract. Instead, he has an identity card and is covered by general human rights and civil rights. If he were truly a cyborg (after Jotterand, 2008b), then the answer to the question of the moral status of such composite beings is already with us, and simply is: they are human. Many of the visions associated with enhancement might be normalized in this way when concrete cases and people are involved. Interestingly, the suspicion of the emergence of a new species between humans and technology was raised recently from the other side: in the ongoing debate on digitalization and artificial intelligence, some people have started thinking about assigning some rights to robots, thus assuming them to be more than mere technology (Chap. 8).

(6) *Human nature*

In view of far-ranging visions related to strong human enhancement (Sec. 7.2), there are references to human nature as a factor limiting or even hindering any enhancement. May humans as they are, whether created by a creator god or the result of evolution, liberate themselves from this connection? Are humans allowed to give up their naturalness in favor of an artificial future created by themselves? The fear of violating human nature, or "playing God" (Sec. 4.3), plays an important role in public debate. The situation is complex however, in an anthropological as well as an ethical regard. Humans are not simply natural but at least also cultural. Plessner (1928) speaks, for example, of man's (*sic*) "natural artificiality" and describes "man" as being a natural being and a cultural being at the same time. No unambiguous argument regarding human enhancement can thus be gained from his perspective on human nature (Siep, 2005; Clausen, 2009): Enhancement is unnatural because it transcends the present natural state. On the other, however, it is also natural for humans not to stop at any level they have reached:

> While some authors [...] declare human "nature" to be sacrosanct, others are of the opinion that human striving for perfection and self-transcendence are part of being human and are therefore to a certain extent actually "natural" (Jömann and Ach, 2006, 35).

The appeal to human nature can thus be employed in any manner desired, i.e., simultaneously both for and against human enhancement. It therefore cannot contribute anything to this issue. In a nutshell, "Human nature is thus contingent" (Clausen, 2009; cp. also Grunwald *et al.*, 2002).

Another argument for rejecting the justification for employing the argument of human naturalness operates with the naturalistic fallacy (Frankena, 1939). It would, according to these authors, be a naïve naturalistic fallacy to argue that we may not technically enhance humans' evolutionarily acquired abilities to see, hear, think, etc. solely because they have arisen naturally and developed through evolution. The fact that we humans happen, for example, to have eyes that only work in a certain segment of the electromagnetic spectrum does not have any immediate normative significance: "Behavioral norms do not follow immediately from evolutionary facts" (Siep, 2006, 312). Some authors claim that limiting human capacities to the naturally provided organic attributes would be an arbitrary museification of humans and would mask the cultural side of being human, part of which is to transcend oneself, i.e., to think and develop beyond what already exists (Clausen, 2009).

In a certain sense it is possible to use human nature as an argument, if not a very specific one. The species *Homo sapiens* has developed by evolution in the course of millions of years. It appears per se suspiciously risky to intervene in this slowly growing nature to any significant degree and on a very short time scale, compared to evolutionary periods. If human enhancement were to take place very significantly and very quickly, we might lose the manner in which we recognize each other to be *humans*, a scenario that would surely cause anxiety. The result of these considerations is, however, only some general obligations to be cautious and to give preference to incremental and reversible steps instead of immediately undertaking a radical transformation of human bodies and minds – which, however, is far beyond the possibilities of science and technology.

The concern that technical enhancement might endanger or even destroy human nature thus does not pose a strong argument at all. Human nature "functions as an entity reflecting anthropological processes of self-clarification" (Clausen, 2009). In other words, it is worthwhile to discuss and quarrel about this concern, but we cannot expect it to provide any clear answers to the ethical questions raised by human enhancement.

7.3.3   Balance of the Ethical Debate So Far

In the ethical debate, many and heterogeneous issues have been raised and discussed. Most have addressed the individual level, considering individual humans with envisaged enhancements (for the level of society, cp. Sec. 7.4). Overseeing the ethical debate, my clear assessment is: so far no *strong* arguments have been raised against human enhancement by technology. This statement must be explained.

"Strong arguments" refer to arguments that build on assumed consequences that will *necessarily* follow the introduction of enhancement technology, without being dependent on uncertain developments in the future. An example of an argument claiming to be strong in this sense is Habermas' argument against genetic interventions in the human germline (Habermas, 2005; cp. Sec. 6.3). His line of thought is that an emerging person made subject to such an intervention during its embryonic stage of development would be instrumentalized without any opportunity to defend itself or even free itself of this instrumentalization in later life. This situation would undermine the *conditions of morality* in general. The core of this argument is that, in this case, human enhancement would be applied to humans unable to give their consent, who would not be able to remove these interventions in their later life. If only responsible persons, after receiving information about the risks and providing their informed consent, were to undergo an enhancement, this risk could be avoided. Responsible individuals, on the contrary, would possibly – at least according to the opinion of the promoters of the converging technologies – gain autonomy by being able, according to their intentions, to liberate themselves from naturally given conditions. Informed consent takes on a central role in ethical argumentation.

However, as long as enhancement measures are only considered for persons able to give informed consent, no space for such strong arguments is available because of the extremely high position given to personal autonomy for moral issues in modern times. Thus, the ethical debate has focused on unintended side effects and risks, e.g., on questions of distributive justice (see above), in a consequentialist manner (Sec. 3.3.2). Arguments of this type are, in terms of validity, only *weak* arguments, which can be seen nicely by looking at the "what if" mode of argumentation frequently used (Nordmann, 2007a). Such arguments have to make a series of more or less uncertain assumptions about the future. Generally, arguments based on the possible emergence of unintended side effects do not constitute arguments *per se* against any technology. Instead, they can be read as imperatives that something must be done politically and socially, perhaps also technically, to limit, prevent, or compensate for the emergence of these unintended consequences. Their purpose is to think about possible problems at an early stage so that it is possible to think about avoiding or solving them just as early. This will keep society from being surprised by the consequences. This thinking is thus part of a precautionary and reflected approach (Grunwald, 2019a), rather than being able to deliver strong ethical arguments.

Viewed from this perspective, in the absence of strong ethical arguments and recalling the inappropriateness of applying human nature arguments, it seems plausible for enhancement technologies to be introduced according to a market model. It is not only possible but even probable that there might be an emerging demand for enhancement technologies caused, e.g., by new lifestyles or increasing competition, as has been explained by various developments:

> [...] several market pressures leading to rapid development of HE [human enhancement] technologies: 1) global competitiveness; 2) brain drain/depopulation economics; 3) national security concerns; and 4) quality of life/consumer life-style demands (Williams and Frankel, 2006, 3).

The situation is similar to the use of weak enhancement regarding doping in various areas (e.g., sports, beauty, and sexuality). A society that has instituted the idea of competition as its central motor at nearly every level, from the economy to the military to lifestyle, will be confronted by ongoing efforts to achieve enhancement, be it by technical or other means. Phrased differently, competition and enhancement belong inextricably together. The pressure of competition plausibly results in measures that first motivate weak enhancement, e.g., doping. The combination of ongoing strong competition and further developments of existing weak enhancement due to the technological imperative and advance means that the development will tend to go beyond doping and lead in the direction of strong enhancement. In a liberal market model, regulation would be limited to compensation for the side effects of a market failure, e.g., for the clarification of liability issues, and to ensuring distributive justice and access:

> Governments may also need to protect the level playing field for consumers, so that the already-enhanced do not act, through non-market means, to protect their status by preventing others from becoming enhanced (Sarewitz and Karas, 2006, 10).

With respect to certain standpoints, this market-oriented scenario may seem ethically problematic or even scandalous, especially considering the situation that the possible human enhancements often meet with uneasiness or even substantial anxiety. This anxiety certainly plays a role as a *fact* and must be taken seriously. In consequentialism, this fact is taken into consideration ethically by including moral anxiety as a cost in the overall evaluation of a technology, possibly resulting in its negative balance. From the perspective of deontology or discourse ethics, however, factual anxiety does not have any normative consequences *per se*. For example, anxiety and rejection could simply result from a lack of familiarity or of familiarization with human enhancement. In this case anxiety evidently could not be attributed any ethical relevance, such as an argumentative role in justification debates. Yet the empirically observable anxiety poses a challenge to ethics, namely as a motivation to search for patterns of argumentation that are possibly hidden behind this anxiety, related to the imperative to make them explicit.

7.4 Are We Heading Toward an "Enhancement Society"?

The emergence of a market for enhancement technologies seems plausible in the absence of strong ethical arguments and according to emerging demands. The market model means that enhancement technologies could be offered as a kind of service analogously to cosmetic surgery today. Consumers willing to enhance themselves would be informed about the available enhancement services, their costs, implications, and possible risks. Then they could decide within the framework of informed consent. This scenario expresses the libertarian ideal of autonomous persons deciding about their personal issues (e.g., Greely *et al.*, 2008).

However, the libertarian perspective might not cover the whole story. It relies on the assumption of free and autonomous individuals. The question is, however, whether individuals are really autonomous, or whether they are subject to external pressures and forces, and/or whether they have a voice in determining the regulating normative systems they have to observe. If we look at possible driving forces of an emerging marketplace for enhancement technologies, we become aware of the crucial role of competition. This strongly resembles that of competition within the world of sports: the stronger the competition and the corresponding pressure of the system on the individual competitors, the greater the willingness of individuals to use drugs for doping might be. Their freedom for self-application of weak or even strong enhancements is limited by external pressure.

A society that has accepted the idea of competition as its central motor at nearly every level, from the economy to the military to lifestyle, might feel a need to achieve human enhancement if this were available. Competition and enhancement are deeply interwoven (Grunwald, 2012). Consequently, the question arises: Are we witnessing a historical change from a performance society to an enhancement society, with an inherent and infinite spiral of initially weak and later strong enhancement including, as critics assume, increased self-exploitation and self-instrumentalization of humans feeling under pressure from competition? Coenen and colleagues point out:

> One could argue that there is growing evidence for the hypothesis that we are witnessing at the moment a transition from a performance-oriented society, in which the fulfilment of predefined tasks is rewarded, to a performance-enhancing society, in which tasks in work life, and even private life, are ever harder to calculate and foresee, and therefore the most pressing task for individuals is the competitive improvement of bodily preconditions and requirements for successful performance (Coenen *et al.*, 2009, 45).

Thoughts of this kind have been expressed in only a few publications to date (e.g., Wolbring, 2008c; Coenen *et al.*, 2009); and most of the reflections on human enhancement technologies refer to ethical questions and criteria at the individual level. While these explorations and reflections have their own value (Grunwald, 2010), it might be that they only address parts of the entire picture. The main motivation of this section is to look for possible other relevant issues mostly overlooked so far. In particular, the focus is whether the debate on human enhancement might tell us something about our society and its current perceptions, attitudes and concerns. The hypothesis is that we can learn from the ongoing debate on human enhancement far beyond consequentialist reasoning (Sec. 3.3.2);

about ourselves today rather than about the future of human nature. While the philosophical debate focuses on ethical aspects of possible and partially speculative *future* developments (e.g., Siep, 2006), it is assumed in other parts of the debate that the acceptance of enhancement techniques and even their use are already widespread:

> Today, on university campuses around the world, students are striking deals to buy and sell prescription drugs such as Adderall and Ritalin – not to get high but to get higher grades, to provide an edge over their fellow students or to increase in some measurable way their capacity for learning (Greely *et al.*, 2008, 702).

Opponents of human enhancement frequently acknowledge that enhancement is already happening. Michael Sandel (2007), for instance, analyzes the mainstream behavior of the contemporary US middle class. He concludes that there is a widespread tendency, particularly with regard to education, training, and child rearing, that displays aspects of enhancement and, in his words, the desire for perfection. Many parents employ all available means in order to maximize their children's abilities. He writes of parents who are willing to pay tens of thousands of dollars in order to increase their child's height in the hopes that they will become a better baseball player, without any guarantee of success. We also see acceptance and willingness to use performance enhancing drugs in the workplace (DAK, 2009) and in the university (see the quote by Greely *et al.*, 2008, above). In the absence of strong ethical counterarguments, enhancement could spread according to market rules. An attractive world market for enhancement technologies and procedures could emerge, such as the one that already exists in plastic surgery. In particular, in view of the fact that Western societies are aging, enhancement could be an appropriate means to maintain competitive advantages for companies and economies. The question for social self-diagnostics is whether we stand at a transition from a capitalist achievement society to an enhancement society. Consider the following:

> In a political-analytical and sociological line of reasoning of this kind, social structures are considered which favour the spread of HET [human enhancement technologies] as are new tendencies which may boost their use. The pathologisation and medicalisation of more and

more emotional and physical states, the commodification of the human body, its use as an improvable tool for competition, and the prospects of radically changing the human body by means of secondstage HET are only some of the aspects relevant here (Coenen *et al.*, 2009, 44).

The changes and disruptions in the world of employment, e.g., as a result of digitalization and automation, presumably have changed our understanding of performance. It often no longer appears sufficient for an individual to perform in a context defined by salaried employment. More and more people feel compelled to constantly improve their performances and possible actions in an increasingly flexible world of employment as well as in other areas of life. Rankings and ratings, regular evaluations, the necessity of presenting oneself and beating competitors are ubiquitous. They apply to the fields of work, of education, to contexts of one's love life, to attempts to become the next top model or some other contest, to use some crazy idea to get the media's attention or to be featured in the Guinness Book of Records. Competition is ubiquitous, and competitiveness is measured by the abilities of individuals. In this way, culture becomes an aspect of competitiveness, education becomes an article of competition, and soft skills are required to increase one's own competitiveness.

Lifelong learning comprises one element in this ongoing challenge, for which traditional education could be supplemented or replaced by human enhancements in the future. An above-average and constantly improving performance at work, a beautiful and strong body, high resistance to stress: these abilities (Wolbring, 2008c) are moving to the top of the agenda for many people. They are expressions of an atmosphere where constant enhancement is necessary just to maintain one's quality of life, not to mention improving it. Competition and abilities are inseparable; thus, the improvement of one's abilities becomes part of the dynamics of development under the pressures of competition. This is because every success in competition only helps for a limited period of time and is constantly threatened by the possibility that others might catch up or pull ahead. Enhancement as an infinite process without a *telos* (Sec. 5.2), which opens up an infinite spiral of further and further improvement. Therefore, the ideas of competition and of human enhancement converge in a synergetic and mutually reinforcing manner.

At the level of political philosophy, competition is deeply related to political libertarianism. Grounding ethical statements on the idea of informed consent with regard to an individualistic approach, which assumes complete freedom and autonomy at the individual level, is a door opener for human enhancement:

> A more mundane vision in a similar vein, presents us with a society in which "morphological freedom" and "cognitive liberty" are core values [...]. In such a society, every individual would have the right to treat his or her own body as fully malleable object. While some critics have denounced this as a reduction of the body to a commodity or a fashion accessory [...], the promoters of morphological freedom argue for a new notion of individual freedom which allows for aesthetic self-realisation, overcoming the "genetic lottery," free experimentation with all kinds of mood- and mind-altering drugs and technologies, and transcendence of the merely human (Coenen *et al.*, 2009, 44).

In a libertarian market model, regulation would be limited to compensating possible side effects of a market failure, assuming, however, that an ideal market without regulation would avoid side effects. Consequently, there are, through a combination of libertarianism, ubiquitous competition, and the imperative of engineering always seeking improvement, strong forces supporting the development and implementation of human enhancement in the absence of state interventions. These forces are part of our contemporary social reality and are already at work. But what about the reality of human enhancement so far? Does it work?

While people often reject the general idea of human enhancement for principled reasons, they seem to be willing to accept and welcome enhancement technologies in specific contexts. In particular, this holds in situations of high competition where people are afraid of lagging behind their competitors and hope for a boost from enhancement. Stories and visions of cognitive enhancement seem to be able to exert real power and influence in spite of the absence of real effects of cognitive enhancers:

> Even where such HET are non-effective, their widespread use changes social interaction, promotes the establishment of new norms, and

raises a variety of questions such as those relating to distributive or procedural fairness, to the fabrics of society, to the relationships between the self-image of individuals and their social roles, and to social key concepts such as competition, ability and disability, happiness, self-realisation, and individual freedom (Coenen *et al.*, 2009, 44).

At this point, it is pertinent to remind the reader of the hypothesis guiding this section: what we can learn from the ongoing debate on human enhancement *is about our contemporary selves rather than about the future*. In order to understand the intensity of the debate and the great controversies involved, as well as the high engagement of many researchers and thinkers, it seems to be most promising to look not at what the various future visions of human enhancement tell us but at diagnoses and perceptions today. A direct pathway leads from the high value assigned to competition in many areas of human life, over the principles of the capitalist economy and its libertarian background philosophy, to human enhancement. From this perspective, the debate on human enhancement is not simply about ethical issues, which could be answered in one direction or another, but also about the type of society in which we live and its implications. The politically explosive nature of this question is obvious. It includes the question: Do we approach a new type of society characterized by crude social Darwinism and the possible end of the welfare state based on solidarity according to the Western European Model?

Again, the world of sports can serve as an illustration. Analogies are often misleading, but it seems promising to take a look at this world from the perspective of doping and enhancement, as exemplified in a recent TA report to the German Bundestag:

> The principal social and political relevance of the topic "Enhancement" arises not because enhancement is perceived as contributing towards a scientifically and technically based "improvement of human beings," but rather because pharmacological interventions to improve performance form part of the "medicalization of a performance (enhancement)-oriented society." The social and political debate about this issue should therefore focus on the likely future status of pharmacological and other (bio)medical strategies and measures for coping with performance targets and demands in a globalized

educational and working environment, and on the consequences of demographic change. To this end, rather than assuming at the outset that adoption of strategies designed to maximize individual and collective performance is inevitable, we need to look into conditions in secondary and tertiary education and at the workplace, and where appropriate adjust performance indicators. Commercial and economic considerations also favor such an approach, at least in the medium and long term. In this regard the example of doping in sport shows how a system of competition could potentially self-destruct as a result of unlimited expectation of ever-improving performance (Sauter and Gerlinger *et al.*, 2011).

Independent of whether this analogy might shed light on the ongoing debate on enhancement in the immanence of the present, we should be aware that stories about the future, such as visions, can exert real power and influence on the respective present (Grunwald, 2007):

In foresight and technology assessment as in any other reflection on the future of HET, one has to be aware of the fine line between taking a broad look at the future and feeding the hype. As has been shown for nanotechnology (Paschen *et al.*, 2004), the enthusiasm which optimistic futuristic visions can evoke is often deliberately utilised as a means of promoting technology development. Such a strategy of "hype and hope" always appears to be precarious. This strategy can have both positive effects (e.g. incentives for young scientists, or arousing and sustaining political and business interest) and adverse effects as well. Among the latter is the danger that expectations will be set too high, making disappointment inevitable. Second, it may popularise the reverse of the optimistic futurism – a pessimistic futurism involving apocalyptic fears and visions of horror, which itself is being increasingly used to raise attention for nascent fields. Third, the focus on far-reaching visions may hinder a sober discussion of the potentials of technologies (Coenen *et al.*, 2009, 38).

Perhaps the most relevant challenge today is not a possibly-arriving enhancement society with problems of access and justice, where cyborgs enter our life-worlds, but more that many people today believe in it, or are afraid of it – possibly because of the aforementioned uneasiness with increasing and omnipresent competition. These thoughts, obviously, are qualitative and speculative. They bring together issues of the current debate on human enhancement and observations of current society with an analysis of what we can learn by investigating ongoing debates on possible futures which include human enhancement. These thoughts need an in-depth analysis and the development of indicators of how the emergence of an "enhancement society" could be measured. The results of those "measurements" could then be used to underpin or to falsify the hypothesis by empirical research.

In any case, doubts should be taken seriously as to whether the dominant, purely individualistic approach of libertarian ethics is appropriate for assessing the full story of human enhancement. If my diagnosis of deep-ranging relations between the debate on enhancement, economic competition, technological progress, capitalism, philosophical libertarianism, and the uneasiness in large parts of the population with regard to increasing self-exploitation are correct, even to a limited degree, then statements of this type may be naïve, because they ignore the fact that individuals are not free but subject to pressure and external forces:

> Like all new technologies, cognitive enhancement can be used well or poorly. We should welcome new methods of improving our brain function. In a world in which human workspans and lifespans are increasing, cognitive enhancement tools – including the pharmacological – will be increasingly useful for improved quality of life and extended work productivity (Greely *et al.*, 2008, 705).

Or statements of this type might be ideological – which is not much better – because they intentionally ignore those pressures and forces. These reflections not only give rise to a focus on the individual aspects of human enhancement, thereby relying on speculative assumptions in a consequentialist manner (Nordmann, 2007a), but also look at what the enhancement debate may tell us about possible deep-lying convictions and challenges of contemporary society and the economy. Concurrently, the current hype on the future of humans in an increasingly digitalized society leads to similar questions and challenges (Chap. 8).

7.5   Responsible Human Enhancement

Responsibilities around interventions into the human body and mind have been established over decades in the medical ethos of physicians and care personnel. These existing assignments of responsibility are manifested in professional roles, in rules of accountability and liability, in regulations and codes of conduct, in relationships such as between doctor and patient, in principles such as informed consent, and so forth. Their entirety forms a normative framework deeply rooted in human practice and culture. Even if interventions in the areas of weak and strong enhancement are, as was shown in the previous sections, not covered by this framework, it could be used as a point of departure for the search for normative frameworks also governing enhancement practices. In the following, we will briefly consider the overarching responsibility issues of human enhancement (Sec. 7.5.1), taking neuro-electric interfaces as an illustrative example (Sec. 7.5.2).

7.5.1   Responsibility Constellation

Most of the more dramatic stories about human enhancement are based on extrapolations of current knowledge and speculative visions of what might come to pass in the future. While technological progress is assumed to be highly dynamic, often expecting even further acceleration, society is regarded as more or less static, without huge changes. This unequal treatment generally leads to an anachronistic situation with a reputed potential for paradise but also catastrophe (Grunwald, 2007). Society obviously would hardly have the capacity to deal with cyborgs, or people who are technically completely transformed or extremely old, if they were to appear suddenly in great numbers and without any advance warning. But the inevitable consequence of this line of thought is either an excessive demand for responsibility in such a large dimension that no viable solution seems possible, or alternatively adverse attitudes and rejection.

This line of argumentation is, however, a simple fallacy. Society is not static but rather is highly able to embed new technologies by stepwise processes of learning and enculturation. The history of the automobile demonstrates these many learning effects with respect to, for example, regulation, establishing drivers' licenses and traffic courts, liability schemes provided by insurance companies, and emergency services taking care of accidents. These elements were not available in the early days of the automobile but were invented and established over time. Similarly, we can expect that society will develop appropriate measures, regulations, and rules for human enhancement over time. The question is thus about responsibility assignments *today* to develop human enhancement and its embodiment into social practices in a responsible manner.

The task today is not to decide about the future of humans and their nature, resulting in responsibilities for their shape fifty or more years ahead. Customarily, human enhancements come gradually. Although the possibility that there might be disruptive leaps from today's "normal" humans to technically completely modified humans and cyborgs cannot be completely excluded, this is only true in the fundamental sense that almost nothing can be excluded with regard to the future. Such disruptive leaps are profoundly implausible, especially considering the complexity of the enhancement measures involved, the difficulty in controlling all of the relevant processes, the related high vulnerability to risk, and the need for learning from the preceding steps.

It is plausible, however, that there will be weak and strong enhancements of humans in the future, in particular because of the absence of strong ethical arguments against enhancement (Sec. 7.3.3), but also because of imaginable demands and because of the gradual steps from healing through weak to strong enhancements. But technically enhanced humans, or human–machine composites, will not appear suddenly. The gradual nature of the emergence of human enhancement reduces the challenge of establishing normative uncertainties for dealing with futurist stories, as well as assigning responsibilities in this respect. By considering both these challenges in the context of a timeline of decades, they lose their fright, or at least some of it. We have time to reflect on the emerging possibilities of human enhancement. Therefore, any responsibility assignments only need to cover today's actors in the field, complemented by careful monitoring to become able to adapt responsibility assignments over time according to new developments as they unfold.

In view of this conceptual situation, we can establish the following theses on responsibility in this field, which require further reflection and concretization, structured according to the groups of different actors involved in the framework of the EEE model of responsibility (Sec. 3.3.1):

- *Converging technologies and related scientists*: The responsibility of the participating scientists (e.g., in computer science, biotechnology, and cognitive science) comprises the usual facets of scientific responsibility (Lenk, 1992) with regard to the goals, research practice, and early reflection on unintended consequences according to standards of good ethical practice (Sec. 3.3). In addition, providing information at an early time and participating in the public dialogue belong to this responsibility.

- *Medicine*: The established ethos of healing has to be extended to emerging forms of enhancements as far as possible and plausible. Existing fields of weak enhancement, such as cosmetic surgery and doping in sports, have contributed to doubts as to whether this ethos alone can provide orientation. Probably, new types of responsibility will have to be explored and established. In particular, human enhancements will modify the roles of physicians: the future physician as body designer? Reflection on the profession and the roles of physicians is needed to help shape this process in a responsible manner.

- *Patients*: The roles and rights of patients are part of the existing normative framework of the medical ethos. People undergoing enhancement, however, are not patients looking for healing. They are more like customers buying enhancement services. Reflecting this emerging role can make use of experiences in cosmetic surgery, as a first step, e.g., for assigning rights and duties.

- *Scientific policy advice*: Since enabling human enhancement can bring considerable social changes, policy advice has the responsibility not simply to consult political principles but to contribute to the social dialogue by spreading its results on a wider scale, e.g., in the framework of technology assessment (Grunwald, 2019a).

- *Research funding*: The funding of research influences the course of science. In determining the topics for funding programs on human enhancement, attention must be paid to empower ethically reflected debate on the topics selected.

- *Ethicists*: Ethical reflection is required both as academic expertise (Sec. 7.3.3) and as contribution to public debate.
- *Social sciences*: Beyond its individual context, the political and social aspects of human enhancement must be examined, e.g., for empirical research in a possibly emerging enhancement society (Sec. 7.4) and the role of competition that is inherent in the capitalist economy.
- *Media*: Differentiated presentations can be expected from the media, with regard to both ethical issues and the object of research, thereby supporting public ethical debate.
- *Society*: Finally, society as a whole is confronted with new questions or new inquiries regarding lifestyles and regulation. The debate about human enhancement could be the occasion for individuals to reflect on the handling of their own body and their perspective toward it.
- *Humankind*: Human enhancement raises questions about the relationship between humans and technology, including in controversies with contested movements such as post- and transhumanism (Jasanoff, 2016; Sorgner, 2016). The debate about human enhancement goes beyond ethical issues by affecting, for example, the self-image of humans, the relationship between humans and technology, the relationship between humans and nature, and the future of humans in light of new technical opportunities such as digitalization (Chap. 8). Conducting debates around these issues in a reflected and well-informed but also inter-cultural manner is required, assigning responsibilities to philosophers, historians, journalists, natural scientists, psychologists, etc.

Ideas about human enhancement extend humans' options for thinking and acting. They raise doubts about what until now went unquestioned, for example, that humans have no alternative but to rely on the customary functions of the bodies they have been born with. The transformation of something that had to be accepted as given into something that could be manipulated is the hallmark of any technical progress. In the degree to which human power to exert control is increased, new room is opened for visions and shaping, but at the same time, as side effects, so are the challenges to compensate for the loss of traditions by finding new forms of orientation. Considering the visionary nature of the prospects of, in particular, strong human enhancement and the long periods of time envisaged for its realization, then in all probability there is sufficient time for us to grapple with the issues that are raised. In the case of human enhancement, there is a good chance that ethical reflection and social debate will not come too late but can critically accompany the development of science and technology. Specific responsibilities have to be determined and assigned at the occasion of specific steps to be undertaken, both in research and development and also in creating business models and establishing companies offering enhancement services and technologies. For each specific case, the above-mentioned list could be used as a point of departure but must clarify all the responsibilities and actors involved in order to make responsibility assignments operable and impactful for further development.

7.5.2   The Case of Neuro-Electric Enhancement

Neuro-electric brain–computer interfaces (Stieglitz, 2006) will provide new opportunities to access the brain. Data exchange between computers and brains can happen in both directions. Computers can capture, interpret, and use the electrical activity of nerves or sections of the brain in order, for example, to control an external device such as a computer mouse or an exoskeleton. And external electrical signals can be applied in order to stimulate certain nerves, such as causing a muscle to contract, for example, to restitute lost or defective body functions (Hildt and Engels, 2009). Mostly, their goal is to enable new possibilities for disabled people to communicate, to interact with their environment, and thus to contribute to their well-being (Campbell and Stramondo, 2017):

> [...], using neuro-prosthetic technologies should actually shape and modify brain processing to improve control of each prosthetic technology and facilitate the ability of paralyzed individuals to convert their motor intentions into purposeful communication and interaction with the world (Scott, 2006, 141).

Technically, electrical brain potentials are recorded by appropriate sensors (e.g., EEG) and analyzed by software tools. The results can, e.g., be used to control an external device such as a mouse for using a computer. Via feedback, the user can train their own brain waves and optimize them to improve the functionality of the technical system. Invasive systems make it possible to measure brain activity in certain regions in a more targeted fashion, to interpret them more specifically, and to translate them into action more purposefully. They enable significantly more flexibility and precision of movement than the use of surface electrodes. Such systems are in an experimental state and must be adapted to each user according to the principle of trial and error (Farah *et al.*, 2004).

Among the visionary ideas about the future of neuro-electric interfaces are being able to play piano with an artificial hand, a paraplegic being able to play football again, or implanting a chip containing foreign languages in the brain to be uploaded on demand (Hennen *et al.*, 2007). Some scenarios of the future consider the possibility of being able to use technical chips to compensate for lost brain functions in degenerative illnesses such as Alzheimer's disease. An artificial hippocampus – the natural one receives and processes different sensory impressions, forwards information, and provides the transition from short-term to long-term memory – in the form of a technical implant (brain chip) would replace the nonfunctioning region of the brain. As long as the goal is to heal the conditions described above, or to restitute lost memory or limb function, these developments take place within the context of classic medical ethos, despite their futuristic sounding possibilities. Ideas of human enhancement can be seamlessly attached to these developments because of the gradual transition between healing and enhancement.

Beyond the positive expectations that, e.g., neurological illnesses could be healed, the new possibilities created by neuro-electric interfaces have also led to concerns. Few references to neuro-electric interfaces fail to mention comments about possible moral misgivings, possibilities for misuse, and ethical problems. A frequent theme is that technical measures near the brain are considered to touch on a sensitive issue regarding the traits of an individual's personality (Clausen, 2006). Estimates of this type express concerns that technical operations near or on the brain pose a threat to personality rights, that the traits of one's personality could be endangered, that a person's individuality could be damaged, and that the possible or probable consequences could include violation of human dignity. Privacy could be endangered, and discrimination in providing medical care could occur (Hildt, 2009). Therefore, the many statements calling for ethical limits, or at least for a procedure that is as safe and judicious as possible, are legitimate.

However, mass media reporting about this field frequently makes no clear distinction between visionary speculation, sober descriptions of the measures being conducted at the frontline of clinical experiments, and realistic expectations for the coming years. The public is thus given the impression that far-ranging types of neuro-electric interfaces are already in use or about to be used, without any normative framework having been discussed and established.

Instead of blind anxiety or badly-informed concern, the EEE model of responsibility calls for a reflected and careful consideration of needs for ethical reflection, for considering the epistemological issues of the quality of knowledge about future developments and about the actors to be held accountable and responsible (Sec. 3.3). The objective is to clarify what normative framework is present for the development and application of neuro-electric interfaces and where the existing framework is being challenged by new developments to such a degree that the criteria for the existence of a standard situation in a moral sense (listed in Sec. 3.1) are no longer satisfied.

As long as the goal is to heal an illness or to restore properties of a dysfunctional or destroyed organ, the established normative framework of healing applies (Müller, 2006). The medical ethos, the doctor–patient relationship, patient autonomy in the form of informational self-determination, ethical standards of research and clinical test procedures, providing fair access to medical care services, etc. are all to be considered. Those directly affected are first of all the individuals who use the neuro-electric interfaces, i.e., as understood in this chapter this is primarily users who hope for an improved quality of life. The question here is whether the limitations, risks, and burdens that might arise are in an appropriate relationship to the expected advantages, or whether the dignity of the affected individuals could be violated. In addition to thinking of the individuals who are affected directly, their social environment, such as relatives and friends, but also broader social effects must be taken into consideration.

Within their current state of research and development, neuro-electric interfaces *currently* do not pose a challenge to this normative framework. The anxiety that can sometimes be seen in parts of the public does not correspond to any normative uncertainty, inasmuch as the neuro-electric interfaces taken into consideration either currently exist, or are possible in the foreseeable future to be employed for medical purposes. Indeed, it is not the *current* developments that are considered problematic but those that might be possible in the future. Independent of possible future applications, some of the issues that are already involved today go beyond medical ethics (in part after Hennen *et al.*, 2007; EGE, 2005):

> Because of their location in the body, implants have a greater closeness to their carrier, and this is true not only of implants incorporating neuro-electric interfaces. The question is still open whether and to what degree an implant could thus become an element of the carrier's identity and of his person.

> It can be unclear what the effect of certain interventions via brain–computer interfaces or stimulations will be on the autonomy of actors. A significant feature of brain–computer interfaces is that their functional integration into brain processes could make the autonomy of the actor and his independence of the technical system questionable (Clausen, 2006).

The question as to the relationship between a neuro-implant and its carrier is posed in heightened form when the attribution of actions and their consequences becomes ambiguous. The situation could arise that a person is not entirely responsible for actions that they carried out, because a neuro-implant did not provide the expected performance. Issues of liability could be posed if, by providing certain data, a retina implant induces its carrier to perform an act that leads to an accident. The clarification of such issues of liability could be markedly complicated.

These brief considerations demonstrate two issues: (1) The normative framework of medical ethics and the related assignments of responsibility still cover an overwhelmingly large share of the development and use of neuro-electric implants or prostheses. Talking about possible futurist developments does not render them outdated. (2) New issues beyond medical ethics arise while considering those more speculative and futurist applications. It is valuable and necessary to analyze them carefully in order to prepare society, but also philosophy and ethics, for reflections beyond established normative frameworks (Chap. 9). Nevertheless, it must not be forgotten that responsibility must always be assigned in contemporary time, relative to the state of knowledge and its valid normative frameworks.

7.6   Changing Relations Between Humans and Technology

The debate over the ethics of human enhancement is superimposed by another discourse that is frequently not distinguished from it: about disappearing boundaries between humans and technology and about expectations and anxieties concerning the technicalization of humans.

The idea of an ultimate triumph of *Homo faber*, who, equipped by converging technologies, sets out to manipulate the world atom by atom (NNI, 1999), was born in the nanotechnology debate. This new *Homo faber* believes, as some say, to have everything under control, including the realms of life and society, thereby developing into a *Homo Creator* (Poser, 2016):

> The aim of this metaphysical program is to turn man into a demiurge or, scarcely more modestly, the "engineer of evolutionary processes." [...] This puts him in the position of being the divine maker of the world [...] (Dupuy, 2005).

An unbroken optimistic belief in progress based on science and technology stands at the core of these visions. Many stories which entered the field also told of human enhancement paradise-like futures. Today, these stories mainly are part of the debates on digitalization and artificial intelligence (Chap. 8).

However, positive visions were quickly converted into their opposite. The presumably ultimate triumph of *Homo faber* was re-interpreted as a final Pyrrhic victory. As a result of the renowned contribution by Bill Joy (2000), an apocalyptic dimension was added to the numerous positive expectations for the future. Instead of serving ends set by human beings, the enhanced technologies could make themselves autonomous and take over control of planet Earth, thus leading to the end of humankind.

While many people are concerned about these stories, transhumanists welcome them in an eschatological sense. In their eyes, it is humanity's responsibility to overcome its own deficiencies (Sec. 7.1.3) by creating a technical civilization that, in the final analysis, makes human beings, with their sufficiently well-known deficits, superfluous. A human-made and then independently further evolving civilization should take the place of human civilization (Savulescu and Bostrom, 2009). This postulate would self-evidently be a move to include human beings into technology in the sense of a self-abolition of human civilization.

Many concerns about human enhancement do not directly address ethical issues but rather are centered around the possible technicalization of human nature (Sec. 7.3.2) and the future of humankind. They converge in fears of subsuming humans or even humanity under the realm of technology to an increasing extent. This word collocation contains generalizations in two respects simultaneously (cp. Grunwald and Julliard, 2007, 79):

> A generalized view toward increasing technological stakes in humanity, in which it is no longer a question of individual, particular technologies, which could be reflected and shaped one by one, but of technology as an overwhelming and quickly developing power in general.

> A generalized view of humanity and not of individual human beings and, with it, a generalizing view of the genus *Homo*, or in (philosophical) anthropological respect, of human nature in general.

These perspectives can imply various semantic connotations, especially regarding the distinction between the individual and social sides of human beings. Accordingly, subsuming humanity under technology can result in greater regulation of human beings, both individually and collectively, as is currently discussed in the area of self-driving cars (Maurer *et al.*, 2016). People's apprehensions concerning a possible technicalization are similar: loss of human autonomy and control, possibilities for external manipulation, and instrumentalization of human beings. These concerns might express themselves quite differently, depending on the technology and context under consideration. The following semantic aspects of a possible technicalization of humans can be distinguished (cp. Sec. 2.2 for the understanding of "technology"):

- Implantation of technical artefacts into the human body (e.g., pills, artificial replacement parts, prostheses, monitoring devices) at the individual level. Technicalized humans would simply be human beings with more technology internalized. In this context, ethical aspects such as informed consent or the right to informational self-determination are the focus of the debate (Sec. 7.3.3), and the consequences that this form of technicalization could have on a massive scale would be of less concern.
- Technical organization of society with corresponding effects on the individual (bureaucratization, militarization, monitoring, etc.) would be a technicalization at the level of collective entities. This could lead to a transformation of human society into a technical society regulating everything due to technological efficiency. The endpoint could be a totalitarian society with humans becoming servants of a (socio)technical machinery.
- Technical acquisition of knowledge about the human body and the mind, for example, about genetic dispositions or about physiological processes in the brain. This could lead to the technicalization of the human self-image: if the human body and mind increasingly become the object of technical interpretation and are regarded as a kind of machine, then it will only be a small step to considering humans in general as machines (which is similarly an issue in the debate on digitalization, cp. Chap. 8). Progress in neurophysiology, biotechnology, genetic engineering, and nanotechnology have, in many circles, allowed the idea of a technical human to come much closer, i.e., the description or even re-creation of human beings as machines (Chap. 9).

It is understandable that humans care about the future of humanity. However, this future is neither completely in our hands, nor is reliable knowledge of it achievable. Taking care and reflecting prospectively about possible futures of humankind, ranging between humans taking full control over everything in freedom and autonomy or becoming subject to techno-totalitarianism, must therefore not be undertaken with a prognostic attitude but with regard to developments in contemporary times. In this way, the consequentialist scheme (Fig. 3.1) becomes sensible even in the absence of reliable future knowledge: The reflections can contribute to gaining orientation today on how to proceed with human enhancement.

This prospective reflection can, for example, be directed to the interpretation of humans as, ultimately, a technical system, e.g., a machine (cp. Chap. 2) as a follow-up to their complete naturalization. Enhancement then would be quite consequent, as it is for every technical system. Interpreting humans generally as machines is semantically completely different from the implantation of technical artefacts into the human body. Enriching humans with technical artefacts does not prevent us speaking of human beings as trans-technical entities, i.e., as beings who may profit from the technology in their bodies, but it does not allow us to speak of them as technical beings. The relation of technology to humans is a crucial issue in contemporary human self-understanding (cp. Chap. 9).

Chapter 8

Robots and Artificial Intelligence: Living Technology?

In the preceding chapters, we have mainly considered the advent of new technology, frequently from the field of converging technologies, into the realm of living systems, bacteria and viruses, plants, animals, and humans. In each case, existing living systems served as the point of departure for biology and biotechnologies to investigate and understand them at the micro-level, to modulate them as technically functioning systems, and then to rebuild or replace natural mechanisms with technical ones. This move opened up the pathway for deeper interventions into living systems and for enhancements to realize human purposes, which gives rise to several normative uncertainties, as were explored above.

In this chapter, we will turn the perspective around. We start by investigating recent developments in the field of digitalization and artificial intelligence (AI), which make increasingly autonomous technology possible. As soon as this technology, e.g., an autonomous robot, is able to behave independently in an unknown environment, to learn, and to behave similarly to humans (in some respects), new debates around responsibility, human–machine interfaces, control, and the moral status of the robots themselves emerge, as can already be witnessed. The heading "living technology" thus will be used in a metaphorical manner only (Chap. 2). The question of whether and in

Living Technology: Philosophy and Ethics at the Crossroads Between Life and Technology
Armin Grunwald
Copyright © 2021 Jenny Stanford Publishing Pte. Ltd.
ISBN 978-981-4877-70-1 (Hardcover), 978-1-003-14711-4 (eBook)
www.jennystanford.com

what respect these technologies could be considered as some form of life in the future is still open.

The narrative of artificial companions, when used analogously to human companions, serves as an illustrative door opener, telling us a lot about current expectations (Sec. 8.1). Many developments and implementations of autonomous technology are ongoing, in fields such as industrial production, autonomous vehicles, and care robots (Sec. 8.2). The resulting major normative uncertainties and ethical issues will be considered (Sec. 8.3), which give rise to reflecting on responsibility in this rapidly ongoing development (Sec. 8.4). The chapter concludes with the question of whether *increasing* the technological autonomy which is made feasible by digitalization and AI could *decrease* human autonomy.

8.1 Artificial Companions[1]

Robots have been familiar to humans for decades. Long before they were available technologically, robots played a central role in science fiction (SF) novels and movies. From the SF literature, and especially from movies in cinemas and dramas on TV, we are used to seeing robots taking over important functions, cooperating and communicating with humans, being empowered and allowed to make decisions on their own, showing emotion, and being able to carry out independent actions. Robots belong to the entirely natural inventory of the imagined future worlds of global blockbusters. Many of these worlds would be hard to imagine without artificial companions or enemies of the human protagonists. Well-known examples are the programmable machine man from Metropolis by Fritz Lang (1927), HAL-9000, who was presumed to be infallible in Stanley Kubrick's 2001: A Space Odyssey (1968) and the ball-shaped R2D2 from Star Wars by George Lucas (1977). A topic of the movie I, Robot by Alex Proyas (2004) is how the NS-5 robot Sonny attains consciousness in an emergent manner. During this long tradition of robots in science fiction, which frequently met with great public success, the social appropriation of robots has already taken place. The assumption has been expressed over and over again that many

---

[1]This section closely follows the respective section in the chapter "Robots: Challenge to the Self-Understanding of Humans," in Grunwald, 2016b, including several quotes.

people would not be very surprised to meet robots on the street, while shopping, or at work (Decker, 2012). A complex and risky process of normalization and enculturation from scratch is no longer required or even possible; it has already taken place. Role models for robots have been prepared and disseminated by SF movies and writings. The question about possible roles for robots in society can start directly from the existing role models in movies, art, and literature.

The size and shape of such robots as companions is usually modeled on humans, as was the case in Fritz Lang's movie (1927), where the machine model perfectly emulates a woman and looks accordingly. As a result of this design, robots have been assimilated to human appearance even in their construction, most effectively in several Japanese humanoid robots. For many practical purposes, this is advantageous, if not technically necessary. Robots must be similar to humans in size and shape if they are to be used, for example, as an assistant or companion (Böhle and Bopp, 2014) in the environment where humans live, e.g., in a kitchen, a care home or a smart factory. The shape of robots visually demonstrates their closeness to humans and their compatibility with the human life-world.

The role of the robot as an artificial companion for humans has received particular attention (Böhle and Bopp, 2014). Possible and plausible concepts for the relationships between humans and robots are available using the rhetoric of assistance, of the colleague, and of cooperation, such as in the envisaged digitalized industrial production (Industry 4.0; cp. Sec. 8.2.1). These concepts are propagated at the level of R&D policy and by related research projects, e.g., by the aims of the European Commission's ICT policy (according to Grunwald, 2016b):

> We want artificial systems to allow for rich interactions using all senses and for communication in natural language and using gestures. They should be able to adapt autonomously to environmental constraints and to user needs, intentions, and emotions (EC, 2012, 12).

Reaching these goals obviously needs substantial effort and financial resources for research and development. The European Commission formulated more specific research goals as follows:

[...] unveiling the secrets underlying the embodied perception, cognition, and emotion of natural sentient systems and using this knowledge to build robot companions based on complexity, morphological computation, and sentience [...] (EC 2013, 168).

The technical language used to describe humans is probably most interesting in this quote. Humans are characterized as "natural sentient systems" while the aim in developing robot companions is to give them properties and capabilities similar to humans, including sentience. In the German long-term project "A companion technology for cognitive technical systems," funded by the German National Science Foundation (DFG), the vision reads as follows (see also Dautenhahn, 2007):

Technical systems of the future are companion-systems – cognitive technical systems, with their functionality completely individually adapted to each user: They are geared to his abilities, preferences, requirements and current needs, and they reflect his situation and emotional state. They are always available, cooperative and trustworthy, and interact with their users as competent and cooperative service partners (Wendemuth and Biundo, 2012, 89).

Accordingly, such robotic systems should be designed to act as perfect companions, completely adapting to the needs of their owners, as a kind of ideal servant, being "always available, cooperative and trustworthy" (quote above). Roles for such artificial companions have been differentiated as follows (cp. Böhle and Bopp, 2014):

- *Artificial companions as guardians* "should accompany and supervise the user while monitoring his or her health status and environmental indicators (e.g., room temperature, pollution)" (Böhle and Bopp, 2014, 162). They could have a role in ambient assisted living (AAL) for elderly or disabled people in order to protect them against accidents and allow a safer and autonomous life even in older age. In the debate on new assistance functions in automobiles, a discussion is ongoing whether the onboard computer should be able to decide, based upon data received by appropriate sensors, whether the driver factually is able to drive the car responsibly, or is perhaps tired or drunk (Sec. 8.3.2). Assistant robots could supervise humans to improve or ensure their individual well-being or aspects of the common good.

- *Artificial companions as assistants* should enable "the user to fulfil tasks, which she or he would otherwise be unable to perform" (Böhle and Bopp, 2014, 163). The artificial companion should remind the person to, for example, plan an agenda or take medication, could provide the person with helpful information, or could support them in their daily work, e.g., in a kitchen or in an industrial environment (Asfour *et al.*, 2018). The demand placed on companions is, above all, that they be empathetic and socially acceptable (Decker *et al.*, 2011), which requires the development of a corresponding human–machine interface. This type of artificial companion is frequently mentioned in the area of care robots (Sec. 8.2.1).

- *Artificial companions as partners* "appear as conversational vis-à-vis, artificial playmates and interdependent actors. The emphasis shifts from monitoring and assistance to *companionship services*" (Böhle and Bopp 2014, 164). The object of this role is to build relations between humans and robots and to associate emotions with the relationship, in particular for entertainment and giving lonely persons a feeling of having some companionship. The assistance systems Alexa and Siri promise to provide some entertainment functions in this direction, also the robot Pepper (Sec. 8.3.2). The development of sex robots also fits into this category, in a very specific sense.

Humans are familiar with all of these roles from today's life-world and social life. Sometimes it is even possible to name existing job profiles that fit these roles. For example, an artificial companion as assistant could be the perfect butler. Hence, artificial companions are not visions of a future world but expressions of established relations among and between humans, simply transferred to autonomous technology: the role of the guardian, the assistant, and the partner. The very idea of artificial companions is to complement or even replace human companionship. The robot Pepper is among the first realizations of artificial companions:

> Pepper is a friendly humanoid designed to be a companion in the home and help customers at retail stores. It talks, gesticulates, and seems determined to make everyone smile (ROBOTS, 2020).

This expectation implicitly expresses diagnoses, or at least perceptions, that among humans in these different roles, social life and taking care of one another often is not functioning well enough. If the vision of an artificial companion is regarded so positively, then humans apparently are frequently dissatisfied with their present human companions, or they fear that they could be dissatisfied in the near future, e.g., while growing older and becoming more dependent on companionship. A social deficit – the loss of social cohesion, loss of mutual care – could be solved by adopting robot technology which provides artificial companions. Otherwise, the demand for them would not be understandable, in particular the willingness to support their development with public funds. Hermeneutic analysis has to deal with questions about the reasons for the perceived deficits of human companionship, which make this metaphor and its realizations appear so attractive to many.

While the vision of artificial companions is ultimately a rather conservative one rooted in current social life and arrangements, some visions, go much further:

> Imagination is the first step. We dream it, then we do it. Robots have lived rich fictional lives in our books and movies for many years, but bringing them into the real world has proved more challenging. 21st Century Robot has a new approach: A robot built by all of us, tapping into the power of open source software and a worldwide community of talented dreamers. The stuff of science fiction is fast becoming science fact (Johnson, 2015; www.21stcenturyrobot.com).

In this way, future robots serving as artificial companions are ultimately imagined to be the better humans. As companions, they will always be in a good mood and fulfill their role as partner or assistant perfectly; they will be well mannered, and they will not tire of indulging us by serving us. The wish for robots to be better humans, which might be at the core of many expectations of putting artificial companions at our side, is thus a criticism of ourselves which is well known from history (Sec. 7.1.3).

8.2   Autonomous Technologies

For decades, we have observed how technological systems have been equipped with more and more "autonomy."[2] An early application in the mass market was the use of automatic gears in automobiles, which quickly gained acceptance. In the 1980s, the use of industrial robots resulted in the extensive automation of assembly line work, thus making production substantially more efficient but also making millions of jobs superfluous. Robotics is also applied in areas in which human activity is not possible, is unduly dangerous, or is otherwise challenging for human minds and bodies. Familiar examples are the use of robots in space flight, for maintenance in nuclear power plants, or as deep-sea divers.

While many of these developments have taken place in factories, behind safety walls, or in inaccessible areas out of sight of the public, substantial advances in robotics increasingly affect the life-worlds of people. Technical advances in sensors to improve robots' perceptions of the environment, in mechatronics to make movements such as climbing steps possible, and in electronics have been made possible by the enormous increase in processing the available information, which permits robots to take over certain tasks as autonomous systems in a human environment. Window-cleaning robots replace window-cleaning crews, search engines make inquiries on the Internet, autopilots fly planes, service robots work in chemical or nuclear power plants, automated monitoring systems replace or supplement human guards.

Autonomous systems have been a focus of research and technology policy in many countries for decades. They have considerable transformative potential for many scientific and social areas (e.g., mobility, research, industrial production, logistics, medicine and healthcare, security). At the same time, expert and public debates point to a variety of uncertainties, conditions, and risks that the use of these systems could entail, such as the replacement of human labor, loss of control, increasing complexity in the context of machine–human interaction, more complex distribution or even diffusion of responsibility, or the manipulation of information.

---

[2]Speaking of "autonomous technology" is an interesting and ambivalent notion asking for clarification of its meaning, because "autonomy" was used in philosophy as an attribute of humans. Exploring this meaning in juxtaposition and/or comparison with human autonomy will be a task throughout this chapter.

As with other emerging technologies (Grunwald, 2016b), there are currently no broadly shared definitions of what constitutes "autonomous" systems (Gutmann *et al.*, 2015; Moniz, 2015). Paraphrasing different definition approaches from the engineering sciences and computer science, autonomous systems are understood to go beyond modern automation technology in that they should be capable of solving complex tasks in a specific application domain *on their own*. They must be able to interpret situations that cannot be completely described and predicted in advance, develop and adapt their own strategies for action, and make decisions under uncertainties regarding the perception and interpretation of their environment. Autonomous systems should be (report by the US Department of Defence, 2015, quoted after Moniz, 2015, 150):

- self-steering or self-regulating
- self-directed toward a goal
- governed by laws and strategies that direct their behavior

In the same report, autonomy was defined as "a capability (or a set of capabilities) that enables a particular action of a system to be automatic or, within programmed boundaries, 'self-governing'" (Moniz, 2015, 149). The notion of "self-governing" is close to some philosophical perspectives. The word "autonomy" carries specific philosophical implications deeply related to human self-images. Philosopher Immanuel Kant regards humans as able and obliged to act autonomously, i.e., following only rules and regulations which humans themselves have agreed upon and put into force, e.g., in the form of a constitution.[3] Developing, assuring, maintaining, and extending the autonomy of humans has been a crucial element of the European Enlightenment. The scientific and technological advance has been regarded as a major part of extending human autonomy (e.g., Francis Bacon). Against this powerful history of mind, giving "autonomy" to technological artefacts raises questions, in particular whether increasing the autonomy of technology could shrink human autonomy by leading to necessities of human adaptation

---

[3]This determination corresponds to the meaning of the original Greek word "autonomous" where "nomos" means law and rule. It also depicts the central question: can autonomous technology itself determine the rules it follows while fulfilling its functions and performing processes? This question will be a major issue in talking about responsibility assignments to autonomous technology (Sec. 8.4).

or subordination (Sec. 8.3.2). Unlike traditional technologies, autonomous technology can deeply change the relationship between (human) acting subjects and (technical) objects (Sec. 8.5).

8.2.1 Autonomous Technology in Practice Fields

With respect to practice, autonomous technologies are developed for many application fields. While they create normative uncertainties in detail and invoke different ethical issues, they share the overarching challenge of how to assign responsibility in new human–machine configurations (Sec. 8.4). In this section, we will illustrate these issues at the occasion of three major areas of application.

(1) *The future of industrial production*

Industrial production will be digitalized in the context of Industry 4.0 (Sniderman *et al.*, 2016), in accordance with the principles of self-organization (Manzlei *et al.*, 2016). The real-time networking of products, processes, and infrastructures is supposed to significantly change production processes, business models, products and services, as well as the world of work. The organizational concept of Industry 4.0 consists of four basic design principles (according to Hermann *et al.*, 2016):

- *Networking*: Machines, devices, sensors, and human beings can network with each other in order to communicate and exchange data via the Internet of Things and the traditional Internet.
- *Information transparency*: Sensor data expand the information systems of digital factory models to create a virtual image of the real world ("digital twins" of real objects) and enable, e.g., the smart factory.
- *Technical assistance*: AI assistance systems support human beings in making informed decisions and responding to problems more quickly.
- *Decentralized decisions*: Production systems make independent decisions and perform their tasks as autonomously as possible. Only in exceptional cases, e.g., in the event of disruptions or conflicting goals, they will delegate tasks to, e.g., a human supervisor.

The aim is to enable individualized production according to customer requirements, as a radical countermodel to Fordist mass production. The required automation technology is to be based essentially on AI through the introduction of methods of self-optimization, self-configuration, self-diagnosis, and independent cognition. In addition to individualized products, reductions in production times and efficiency gains are to be realized through accumulated learning during production. Geographically distributed production capacities and the respective planning and control systems should cooperate autonomously and thus make better use of existing resources. In realizing these expectations, data processing, learning processes, and self-organization play a crucial role. Human workers and robots should cooperate "on an equal footing," as was formulated around twenty years ago:[4]

> In the future, robots will be far more than special machine tools. They will work together autonomously with human beings to do complex tasks and to do defined work without exact specification of the course of action in unknown or changing surroundings (Steusloff, 2001, 7; translation A.G.).

Shaping autonomous technology is a task confronted with considerable complexity. Technical standards and norms must be developed to enable human–machine or machine–machine communication. Coordination and cooperation between humans and machines require clearly defined interfaces; among other things, it must be clarified who has to adapt to whom (Sec. 8.3.2). Data security and ownership are given high priority and must be legally protected, just as liability issues for the complex allocation of responsibilities between humans and machines must be legally clarified. The resulting comprehensive transformation of work (Börner *et al.*, 2017) does not only involve the necessity of early and life-long education and training initiatives. It also affects the governance and social quality of work, e.g., the future role of trades unions or new employee representation organizations, the international division of labor in a global labor market, and the development of social assurance systems. In light of the expected far-reaching impacts on

---

[4]The quotations given in this section continue to be up to date despite the rapid technological progress that has been made in some fields. Perhaps, they were ahead of their time.

the economy and the world of work, a foresighted and responsible design perspective that includes society (Mainzer, 2016, 222) is urgently needed (Moniz, 2015).

(2) *Self-driving cars: autonomous vehicles*

In recent years, human drivers have been provided with more and more assistance based on advanced sensors, real-time evaluation of the collected data, and actuators implementing conclusions made by algorithms. Processors and sensors are increasingly able to observe the traffic situation in the surroundings of a car in real-time and determine the next steps to be taken in order to adapt the car to the respective traffic conditions. This development has already led to a partial automation of driving in new vehicles. Highly automated systems can autonomously change lanes and exert other functions. Prototypes have already driven millions of kilometers without any human intervention. In some countries, test fields have been set up where highly or fully automated vehicles can operate (cp. EK, 2017, for the distinction between highly automated and autonomous or fully automated cars).

Frequently, autonomous vehicles are regarded as key to future mobility (Maurer *et al.*, 2016). This opens up a wide range of new mobility options beyond traditional individual mobility with a private car: new mobility conceptions and patterns, new business models for mobility providers, and new combinations of private and public transport, or even a blurring of the traditional boundaries between them, could become possible. This property alone makes self-driving cars a possibly disruptive innovation. Despite the many and far-ranging positive expected consequences of autonomous vehicles concerning safety and comfort, certain risks are also implied; some of these are well known from traditional driving, while others are new and often related to the digitalization of driving, which is necessarily involved (Grunwald, 2016d; Grunwald, 2018). Due to technological problems, or in situations for which the technology is not prepared, accidents specific to autonomous vehicles can occur. Autonomous vehicles add new types of *systemic* effects to the existing ones. Through the control software and the reliance on the Internet, new effects could emerge, e.g., with respect to safety and security. Engaged companies in the automotive sector and in IT are confronted with high risks of return on their investments because of the high costs of R&D for autonomous cars. The labor market could be put under pressure, e.g., by job losses for truck drivers. Also, risks of privacy and dependency might occur – these are not new but could be increased. Therefore, many concerns arise around the issue of public acceptance of autonomous vehicles. In addition to the risks mentioned, the so-called ethics dilemma (the trolley problem; cp. Sec. 8.3.1), as well as the unclear distribution of responsibility and liability among all the actors in the field (Sec. 8.4.2) are major issues of public debate (Grunwald, 2018).

Research is also ongoing on the co-evolutionary dynamics between automation and daily mobility patterns and routines. The diffusion of the various incarnations of automated vehicles (AVs) will, depending on how they are expected and designed to act and interact with one other and with humans, increase complexity in road traffic, in particular in mixed traffic where AVs and human drivers co-exist. Since road traffic is the result of a network of permanently negotiated and reordered social relationships that goes far beyond simple rule-obeying behavior, at a certain level of automation, AVs will become social actors within this network (this refers to the issue of technology possibly taking the role of subjects introduced at the beginning of this section). This creates numerous research challenges, including: whether humans will attribute agency to AVs in traffic; whether and how these technologies have to be able to negotiate in certain traffic situations; whether automation-compliant behavior by humans in traffic should become a future regulatory principle; and whether values could be implemented in AVs, and if so, which values and how (Grunwald, 2018). In this open space of possible developments, challenges for the distribution of responsibility emerge (Sec. 8.4).

(3) *Care robots*

Health and care are important fields of application for automated technologies. In light of an increasingly aging society and a growing share of people in the total population who are in need of care, the future of care is a major societal challenge. Autonomously operating service or care robots and assistance technologies, in combination with neurotechnologies (e.g., exoskeletons), are considered to have great potential to support care (Kehl, 2018). AI plays an important role here in enabling technologically autonomous systems to act adequately in complex environments, e.g., through real-time detection of relevant and sometimes rapidly changing framework conditions during active operation, or of the condition of the affected persons, e.g., in the case of variable dementia (Decker *et al.*, 2017). Since humans and technology come into close contact in these fields of application and the affected people are often in need of care and thus might be helpless against potential malfunctions of technology, an appropriate recognition of the person's condition is extremely important.

Intensive development efforts have been taking place for a long time. However, only a few products are used in care practice so far (Kehl, 2018). Mostly, these are special applications (e.g., eating aids, or therapeutic aids such as the robot seal PARO). They show limited autonomy and do not correspond to the common idea of a robot assistant, or even a companion (Sec. 8.1). More complex robot assistants have not yet progressed beyond the status of a research platform. Some activities have even been stopped after substantial funding had been spent because of slow progress, e.g., a Japanese robot line dedicated to the capability to lift humans needing care and turn them in their bed or move them to another bed.

Beyond technical issues, one reason that robotic care applications are still not widespread is that so far, too little attention has been paid to the needs and situations of people requiring care (Kehl, 2018). Many people fear that care work will increasingly be subject to a mechanistic understanding, i.e., that care could be reduced to purpose-related concerns only, under pressure of economic efficiency. It is therefore all the more important to evaluate ethical aspects prior to bringing the technology into use and to integrate these evaluations into a tangible design of applications, following the aim to develop and implement robots to support systems of "good care."

Because individuals and their relatives are directly affected in all fields of health and care, and questions of dignity, autonomy, and humanity arise, it is obvious that ethical and legal questions must be considered when debating "good care" and developing such technologies. The practical needs of the patients, their relatives, and the nursing staff must also be taken into account early in development in order to avoid purely technocratic solutions. Not only is the problem- and addressee-related approach to technology design crucial here but also the question of the right timing (Decker and Fleischer, 2010). Technology design in this field must therefore include the groups of persons involved to a particularly high degree, while responsibilities again will be distributed across various groups and actors.

8.2.2 Autonomous Problem-Solving and Machine Learning

In the fields briefly sketched above, something like "autonomy" must be realized by means of new technology. The success story of autonomous technologies over recent decades – though most of the expectations concerning an Industry 4.0 or care robots are still at a visionary stage – is grounded on huge technological advances in the following areas:

- In the course of miniaturization made possible by microsystems technologies and nanotechnology, the size and costs of sensors and of memory technology have been heavily reduced, thereby drastically increasing the possibilities for acquiring and storing them in extremely small devices. Sensor and storage technology today is often invisible to humans and hidden in the built environment.
- The fast digitalization technically driven by computational power, miniaturization, and new algorithms allows for data-mining (Witten *et al.*, 2011) in huge data sets (Big Data), for pattern recognition and on-time evaluations, as well as for optimization of action strategies.
- Machine learning (this notion fits better than the frequently used AI metaphor, cp. Chap. 9) is frequently based on neural networks which have been trained by applications to known data sets (Alpaydin, 2004). They allow robots to learn "on the job," e.g., from evaluating the many feedback loops performed and statistical calculation.
- While all this happens in the cosmos of the digital parallel worlds of data, models, statistics, calculation, and algorithms, autonomous technology also needs instruments for exerting action in the real world: tools, actuators, access to analogous technologies.

Based on these elements, autonomous technology becomes able to solve problems without having a set of predetermined strategies in store, except perhaps the building bricks of a strategy. The classical application in direct analogy to humans in their life-world is that a robot has to find its way through surroundings unknown to it, for instance, by overcoming or circumventing obstacles when moving forwards. Another example is soccer-playing robots, which requires coordination of the actions of several players. Autonomy here does not mean a robot's *full* capability to determine the rules of its own behavior – which would be the analogue to philosopher Immanuel Kant's notion of human autonomy – but the capability to tackle unprecedented and unprogrammed situations independent of remote control.

This expectation has become reality to some extent. For example, autonomous automobiles (Maurer *et al.*, 2016) can be seen as autonomous robots moving on wheels to transport people or goods in complex road traffic, which is a sequence of continuously unexpected events, in particular in chaotic urban environments. Likewise, drones are flying robots that can autonomously search for their target in an area where enemies, terrorists, or simply lost hikers are suspected to be. Many applications are expected in the field of caregiving, where however, development is slower than expected because of the extremely high complexity of human actions in care (Kehl, 2018).

In speaking about autonomous technology, e.g., robots, terms from the fields of learning and problem-solving are often used. This use of language reaches back to the early AI movement in the 1970s. The roboticists' fundamental assumption about problem-solving goes back to a statement made decades ago but which still expresses their intention and target:

> Solving problems, searching, and planning are the means with which instructions [...] can be obtained [...] out of a large knowledge base [...] it is a matter of defining the initial state, the goal, and any known intermediate states [...]. In general, a number of paths lead to the goal. [...] In planning, a means-end table is first drawn up in order to recognize central decisions (Decker 1997, 12; translation A.G.).

Autonomously problem-solving technologies are based on a similar and rather simple conceptual scheme. They need sensors to receive and recognize data from their environment, capabilities for storing these data, algorithmic power to evaluate them in order to create an appropriate model of the environment and to draw conclusions for actions to be taken, as well as instruments to exert the respective actions. After having performed the actions, the consequences will be monitored by the sensors and compared to expectations and pre-determined goals. Based on these feedback data the next steps will be determined, and so forth. This model applies to bots on the Internet as well as to real world robots, such as the robot Pepper. Many thousands of copies of Pepper are already cooperating with humans (see SoftBank Robotics, n.d.):

> Pepper is the world's first social humanoid robot able to recognize faces and basic human emotions. Pepper was optimized for human interaction and is able to engage with people through conversation and his touch screen.
>
> Pepper is available today for businesses and schools. Over 2,000 companies around the world have adopted Pepper as an assistant to welcome, inform and guide visitors in an innovative way.

Having performed several cybernetic feedback loops, the autonomous system can be described analogously to humans as a system which learns, in a certain sense, from its environment and from the effects of actions taken (Sec. 8.4). Learning "on the job" then consists of a series of learning cycles by which the autonomous system will steadily improve its functionality, effectiveness, and efficiency.

This learning model shows a deep similarity to cybernetic planning (Stachowiak, 1970). This idea of a cybernetic loop, well known from systems theory (Chadwick, 1978) has been used in robotics for decades:

> Learning consists of the reorganization and re-evaluation of the individual links within a neural network [...]. We have previously spoken of supervised learning, by which, for example, human beings exercise control. If we go further to unsupervised learning, then we replace the monitoring system by a set of well-defined rules for learning. The entire system optimizes itself according to these learning rules (Schlachetzki, 1993, 78f.; translation A.G.).

An illustrative example of dealing with obstacles is a courier robot moving around in a building. Through sensor signals, the robot generates a model of its surroundings while it is moving. As long as these surroundings are static, the model produced will consist of walls, doors, elevators, etc. and can be used without problems. During operation, the robot constantly checks, by means of sensor technology, whether its model is still up to date. If a door which is normally open is found to be closed, the robot encounters a breakdown of the plan, just as when an obstacle unexpectedly prevents it moving ahead. Plan breakdowns designate deviations of the real situation from the expectations. In such cases, the robot defines the area in which a difference between the model of the surroundings and reality occurs as a region of interest (ROI) (Kinnebrock, 1996, 77ff.). Through experimental handling of the unexpected situation, the robot can gather experience. It can try to get the obstacle to make way by giving an acoustic warning signal (the obstacle could be a human being who steps aside after the signal); it could try to push the obstacle aside (maybe it is an empty cardboard box); or, if nothing helps, the robot could notify its operator. Maneuvers such as parking or making a U-turn in a corridor can be planned in this manner. One of the most important challenges in this work is recognizing and classifying the plan breakdowns, so that the robot is able to diagnose the right type and take the appropriate measures as quickly as possible. This is a particular challenge in autonomous cars, because of the high diversity and dynamics of real word traffic conditions, e.g., in an urban area.

The heart of the underlying cybernetic planning concept (Stachowiak, 1970; Churchman, 1968; Chadwick, 1978) consists in a cybernetic feedback loop: a planning system plans to change certain parameters of its surroundings and takes measures to achieve this goal. It then monitors the effects of the implementation of the measures and evaluates them against the expectations. Deviations from the expectations are detected by means of this feedback control mechanism and are taken into consideration in subsequent measures. Learning consists in repeated runs of this cybernetic loop, with a corresponding accumulation of empirical information. The same model of cybernetic behavior is often applied to living systems, which also have to adapt to their environment and its changes over time in order to survive. Hence, autonomous technology is enabled to behave similarly to living systems. Autonomous technology does not immediately start living when set in operation; however, it expresses some of the properties which have been characteristic to forms of life so far. This observation is at the core of the investigations in this chapter and gives legitimization to including autonomous technology in this book, which is dedicated to life and life–technology interfaces.

8.3   Ethical Issues

Recent and expected developments toward a more autonomous future technology lead to normative uncertainties and ethical challenges that have completely different roots from the preceding chapters. There, we considered technical intervention into already existing life; we now address new technologies possessing more and more capabilities, which are similar to living entities in some respects. Hence, the origin of the normative uncertainties to be discussed in this section lies in the increasing power of technology *as such* rather than in interventions into living organisms. Consequently, mainly philosophy and ethics of technology have to be addressed, rather than philosophy of life and bioethics. In particular, recent specializations in the ethics of technology need to be included, such as machine ethics and robot ethics (e.g., Anderson *et al.*, 2011; Lin *et al.*, 2012).

Ethical considerations of autonomous technologies go back to the famous rules for robots proposed by Isaac Asimov in 1942 in the realm of SF literature (Asimov, 1950):

- A robot may not injure a human being or, through inaction, allow a human being to come to harm.
- A robot must obey the orders given it by human beings except where such orders would conflict with the First Law.
- A robot must protect its own existence as long as such protection does not conflict with the First or Second Laws.

These rules have been widely discussed. It has been shown that they do not cover the entire set of possible ethical challenges which could emerge out of the behavior of autonomous machines and their interactions with humans, e.g., in debates around autonomous vehicles and new human–machine interfaces. In this section, we will consider ethical dilemma situations (Sec. 8.3.1), and challenges of adaptation and control (Sec. 8.3.2), as well as the question of the moral status of autonomous technology (Sec. 8.3.3).

8.3.1   Ethical Dilemma Situations

Ethical dilemma situations regarding autonomous technology have mostly been discussed at the occasion of self-driving cars (Maurer *et al.*, 2016), in spite of their extremely rare occurrence in practice. These dilemmas are characterized by the fact that an automated vehicle might have to decide which of two evils should be preferred in cases of absent alternatives (JafariNaimi, 2017). Often, they are illustrated by pictures showing an autonomous car driving along a narrow urban road, when a child unexpectedly jumps into the road, and avoiding injuring or killing the child would necessarily cause harm to others, e.g., passengers in a car travelling in the opposite direction, or pedestrians on the other side of the road. These cases are familiar in ethical and legal contexts as the so-called trolley problem (Thomson, 1985). To date, we have considered such situations as tragic occurrences where someone was in the wrong place at the wrong time. However, when it comes to autonomous vehicles, the trolley problem changes its nature. When using autonomously operating machines, tragic constellations in road traffic previously open to blind fate suddenly seem to turn into intentionally programmable situations. When programming an onboard computer, it must be clear how, and according to what criteria, an autonomous vehicle should decide to drive in such circumstances (Goodall, 2014).

A familiar attempt to gain orientation on how the software of the automobile should be prepared to tackle this type of situation is to look for the governing principles. Attempts to categorize and perhaps standardize possible dilemma situations and to design solutions tailored to each problem category would be both helpful and necessary. However, this approach does not work for several reasons. (a) Given the diversity and complexity of the conceivable scenarios, abstract rules such as "damage to property to take precedence over personal injury," or "minimizing harm to humans," run up against the problem that it seems difficult, if even possible, to categorize all potential situations. However, assume that a categorization is developed and proposed. It will then be easy to find a dilemma scenario which is not covered by that categorization, through slight but morally relevant changes in the midst of the infinite space of possibilities among unique dilemma situations. (b) The cybernetic loop as the main learning mechanism of, e.g., autonomous cars (see above) does not help in this case, because the extremely low number of cases renders statistical learning impossible. Furthermore, this loop does not generate its own ethics but simply relies on optimization according to specific criteria. (c) General principles cannot be applied unless a comprehensive assessment of the impact of damage and possible personal injury is available to relate damage scenarios to the behavior of the autonomous car in a dilemma situation. However, this consequential knowledge will not be available because of the high uncertainties involved. (d) The ethical rules for robots proposed by Asimov do not provide orientation here but instead lead back to the same dilemma because weighing up and balancing arguments cannot be reduced to statistical calculation of algorithms.

Summing up, the combination of the infinite space of possible dilemma situations, the unique character of each of them, and the uncertainty or even absence of consequential knowledge render classical approaches inapplicable. This was also the conclusion of the German "Ethics Commission on Automated and Connected Driving":

> Genuine dilemmatic decisions, such as a decision between one human life and another, depend on the actual specific situation, incorporating "unpredictable" behavior by parties affected. They can thus not be clearly standardized, nor can they be programmed such that they are ethically unquestionable (EK, 2017, rule 8).

In this situation, the Ethics Commission simply proposed to optimize technology to keep damage as low as possible, independent of the individual dilemma situation. The Ethics Commission argued that, "a solution that appears plausible from a technological perspective, offers the greatest potential for reducing accidents in most cases and is technologically feasible, is to be preferred to a solution that is not yet feasible given the current state of the art":

Technological systems must be designed to avoid accidents. However, they cannot be standardized to a complex or intuitive assessment of the impacts of an accident in such a way that they can replace or anticipate the decision of a responsible driver with the moral capacity to make correct judgements. It is true that a human driver would be acting unlawfully if he killed a person in an emergency to save the lives of one or more other persons, but he would not necessarily be acting culpably. Such legal judgements, made in retrospect and taking special circumstances into account, cannot readily be transformed into abstract/general *ex ante* appraisals and thus also not into corresponding programming activities (EK, 2017, rule 8).

In particular, this recommendation avoids the necessity to measure the values of different humans and to compare, e.g., the value of younger and older people, of people with different social status, or of the occupants of the self-driving car and other participants in daily road traffic. This meets the fundamental principle of deontological ethics going back, e.g., to philosopher Immanuel Kant, while other ethical argumentations, e.g., in the utilitarian tradition, do allow weighing the comparative value of humans, under certain restrictions. According to the strong significance of the Kantian perspective in Germany and its implementation in the constitution (*Grundgesetz*), it is not surprising that the Ethics Commission was uncompromising in this respect:

In the event of unavoidable accident situations, any distinction based on personal features (age, gender, physical or mental constitution) is strictly prohibited. It is also prohibited to offset victims against one another. General programming to reduce the number of personal injuries may be justifiable. Those parties involved in the generation of mobility risks must not sacrifice non-involved parties (EK, 2017, rule 9).

Frequently, the question is raised in the mass media in the form of a scandal story about whether technology, i.e., algorithms and software, should be entitled to make autonomous decisions about the life and death of humans. This question obviously applies to ethical dilemma situations but may also go beyond them. However, this concern is misleading. In human affairs, "deciding" means considering and weighing up the different options in a given situation and selecting one of them according to reasons rooted in normative criteria. A statistics-based calculation is different from human decision-making by category. The machine does not "decide" but rather calculates and ends up with a result to be implemented (cp. the in-depth consideration in Sec. 8.4.1).

The question remains of who is making the decision, if not the onboard computer (Hevelke and Nida-Rümelin, 2015)? To answer this, we have to trace by whom and how the results of its calculations have been influenced. They are outcomes of the algorithms, the software architecture, the sensors, the evaluation schemes for analyzing the huge amounts of data, the learning capabilities, and the possible restrictions implemented. These were all designed by humans, according to their different roles, and within the framework of different institutions. This view has immediate implications for issues of responsibility (Sec. 8.4.2).

8.3.2 Adaptation and Control

While new technology usually increases the autonomy of humans and opens up more options for making choices, this increase often comes at the price of decreasing autonomy in other respects (Grunwald, 2019a). The availability of new opportunities offered by technology is often accompanied by new forces requiring adaptation. The most famous and probably most significant origin of human adaptation to technology is covered by the approach known as technology determinism (Grunwald, 2019a). Its recent revival occurred at the occasion of digitalization and the predicted new wave of the industrial revolution, in particular with respect to increasingly autonomous technology. In this respect, issues of control frequently enter the debate. If humans have to adapt to technology, will they still be able to control technology, or is there a risk, or even a slippery slope tendency, toward humans becoming servants and slaves of technology (already Anders, 1964)? We will again consider autonomous vehicles as an appropriate illustration of adaptation and control issues in the field of autonomous technology.

The first case is not very dramatic in ethical respects but illustrative. In highly automated driving, it is possible for humans to drive the car but also for the computer to act autonomously, i.e., without any human intervention. In the corresponding transportation system, mixed traffic consisting of cars with human drivers and automated cars must be managed. This situation requires establishing communication patterns between both types of drivers: humans and computers, for example, to negotiate unclear situations. The question arises of what these patterns should look like to ensure the absence of misunderstandings, which could lead to accidents. The Ethics Commission simply recommended:

> To enable efficient, reliable, and secure human–machine communication and prevent overload, the systems must adapt more to human communicative behavior rather than requiring humans to enhance their adaptive capabilities (EK, 2017, rule 17).

While this sounds evident and can easily be prescribed in ethical guidelines, it will be arbitrarily ambitious to realize. Humans have several communication patterns working over the distance between one car and another or between a car and pedestrians. Gesturing with hands in a specific way will probably be understood appropriately by the addressee, e.g., in an unclear situation of who should pass first, when there is an obstacle on the road or in a very narrow passage. It seems difficult to imagine computer technology being able to understand human gesticulations without risk of misunderstandings. This will be similar in digitalized factories (Sec. 8.2.1), when robots and humans cooperate "on an equal footing."

However, a similar challenge arises if computers are able to understand human mimics. According to the state of the art, AI-based autonomous technology has admirable capabilities to recognize human faces, and partially also human emotions, e.g., in the case of the robot Pepper: "Pepper is [...] able to recognize faces and basic human emotions" (SoftBank Robotics, n.d.). Nevertheless, regarding the individuality and diversity of humans and the large range of different contextual situations, it might be difficult to avoid misunderstandings, possibly resulting in adverse effects or even damage. A typical technical solution is to standardize human communication with AI and robots. While this is in contrast to the recommendation of the Ethics Commission mentioned above, it is familiar in human–machine interfaces so far. Usually, humans have to adapt their behavior to the prescriptions determined by the respective software company, e.g., in the field of administrative software in huge institutions. This situation is clearly expressed and illustrated by the need for educational courses for staff members as soon as a new software system, or even a new version, is installed. In the light of this experience, realizing the above-mentioned recommendation of the Ethics Commission would require converting established developments and might thus be difficult to achieve.

An even more pressing ethical issue is the question of control, in particular who should be the master in new constellations of humans co-existing with autonomous machines. According to traditional understandings in ethics and legal regulation, humans should always have final control and should always be able to overrule the computer. This position, which still is part of the Vienna Convention on Road Traffic (https://www.unece.org/fileadmin/ DAM/trans/conventn/Conv_road_traffic_EN.pdf), corresponds to traditional understandings of humans as masters of technology. In § 8,1 it states: "Every moving vehicle or combination of vehicles shall have a driver," followed by "Every driver shall at all times be able to control his vehicle," and "A driver of a vehicle shall at all times minimize any activity other than driving." This picture of the driver/ car relationship contradicts the narratives of autonomous vehicles. Even when this regulation is changed, as is ongoing or completed in some countries, most humans will stick to the position that humans should keep final control and should be able to overrule the machine at any time, because this position corresponds to the idea of human autonomy. However, thought experiments might call this position into some doubt, at least in certain cases. Consider:

- A human driver uses a highly automated car. He or she plans a terrorist attack by driving the car into a group of pedestrians. The car will be pre-programmed to avoid collisions and, in particular, to avoid killing people. Is the claim for individual human autonomy while driving a car so strong that it should allow this human driver to overrule the software in this case and to deliberately use the car as a lethal weapon?
- A human driver takes over control in a highly automated car and navigates the car through daily traffic. Now, the computer observes that the driver is not performing well, maybe drunk or tired, or potentially undergoing, e.g., a heart attack. The computer would be able to take over control and drive the

car safely to a stop. Would it be entitled to overrule individual human autonomy in this case?

In both cases, the answer is clear for ethical reasons: the computer must take over and overrule the human in order to prevent damage to other humans, and/or to the driver. This is quite clear and in no way surprising, because the autonomy of a particular human does not allow for arbitrary action but rather ends in the autonomy and rights of others. However, answering the questions in this direction uncovers an even more interesting issue. In both cases, the computer, supported by its sensors, decides whether the human driver should retain control, or not. This implies that the human mandate for driving a car has been made dependent on the computer's observations and decisions. The computer would become the master of the car and of the driver: permission as to whether a human driver may take control would be given or refused by the algorithm. This configuration, with the computer being the master while the human driver would be required to ask it for permission, would be *permanent*, not confined to rare situations. The computer would always monitor the performance, behavior, and conduct of the driver. Any human control would only be borrowed from the computer.

In terms of consequentialist ethics, this is not a fundamental problem, at least if everything works well, if the computer has clear rules of when to give permission and when to reject the request, and if these rules have been determined in a knowledge-based, ethically reflected, and politically legitimated process. It belongs to human autonomy to renounce parts of that autonomy for good reasons. However, transfers of autonomy from humans to technology should be monitored and well reflected in order to avoid running blindly beyond points of no return or moving into unrecognized and irresponsible dependency on autonomous technology. Hence, this is not an ethical issue but more an issue of anthropology and philosophy of technology (Chap. 9).

A follow-up to this debate around issues of control is concern about an increasing subordination of humans to technology in social life. In a digitalized society, more and more tasks will be performed by automated systems operation with Big Data and algorithms. Software functions similarly to institutions and regulates human behavior (Orwat *et al.*, 2010). Software-based systems could be used to abandon human freedom by applying arguments of safety, or they could sanction undesired behavior by social credit systems, as is currently implemented in China (e.g., Liu, 2019). Software is highly value-laden (Brey, 2009).

Road traffic is again an appropriate example for the techno-based safety argument. Human drivers are responsible for more than 90% of accidents and fatalities in daily traffic. Most accidents are caused because humans do not follow the rules but exceed speed limits, do not observe alcohol prohibitions while driving, are tired or aggressive, and so on. However, onboard computers stick to the rules, do not take drugs or alcohol, and never get either tired or aggressive. Expectations of autonomous vehicles tell stories of largely abolishing accidents and fatalities by an increasing diffusion of self-driving cars. But could this positive expectation lead to the prohibition of humans driving themselves and eventually to the introduction of a perfect technical system which guarantees safety, but without any human freedom, and thus to the emergence of a completely technical world without surprise and exploration, without emotions and space for dreams and utopias, with humans serving the technical infrastructures to make all this possible? Public concerns about autonomous vehicles often give emphasis to this point. The German Ethics Commission reacted with a possibly surprising guideline:

> One manifestation of the autonomy of human beings is that they can also take decisions that are objectively unreasonable, such as a more aggressive driving style or exceeding the advisory speed limit. In this context, it would be incompatible with the concept of the politically mature citizen if the state wanted to create inescapable precepts governing large spheres of life, supposedly in the best interests of the citizen, and nip deviant behavior in the bud by means of social engineering. Despite their indisputable well-meaning purposes, such states of safety, framed in absolute terms, can undermine the foundation of a society based on humanistic and liberal principles. [...] Decisions regarding safety risks and restrictions on freedom must be taken in a process of weighing-up based on democracy and fundamental rights. There is no ethical rule that always places safety before freedom (EK, 2017, 20).

This statement aims to create a barrier against the complete technicalization of society, resisting complete subordination of humans under technological systems. It calls for open spaces of human freedom, even if they are sometimes "objectively unreasonable." The weighting between technologically enforced issues of safety and desired behavior, on the one hand, and the legitimate postulate for human freedom, on the other, has to be balanced over and over again. Regarding the advent of autonomous technology in modern society, it seems plausible that the debate between safety/security and freedom will be of major significance in the future.

8.3.3   The Moral Status of Autonomous Technology

The question of the moral status of technology might sound fuzzy. Technology consists of objects and processes, artefacts and things. In at least Western understanding, things such as cars, computers or production plants do not have any moral status but are just things at the disposal of humans. There is no ethical rule not to torture technologies, or to respect their dignity. Even the question of why those rules do not exist seems fuzzy. So how and why could, or perhaps should, this be different in the case of autonomous technology?

The main source of the question of the moral status of autonomous technology is its capability to make decisions and exert action, e.g., in the case of self-driving cars. Usually, exerting action is related to the assignment of responsibility (Sec. 3.3.1). However, if responsibility was to be assigned to robots in some way, then the question of their moral status would arise, regarding questions of guilt and possible sanctions. We will return to this issue (Sec. 8.4.1). A second source of the question about the moral status of, e.g., robots is grounded in visionary, or (better) speculative expectations. In particular, some neuroscientists, computer scientists, and science writers assume that autonomous technology will develop consciousness and its own intentions, similarly to humans (Reggia, 2013; Chalmers, 2011; Gamez, 2008). In this case, the urgency of the question of their moral status would be self-evident.

As a further motivation for raising the question, I would like to consider the use of language concerning autonomous technologies.

On the Internet, a nice description of the robot Pepper mentioned above can be found:

> Born in Japan in 2014, Pepper is a 4-foot-tall, 62 pound humanoid robot on wheels. Pepper [...] is designed to communicate with people--both as a personal aide and for businesses to use in interactions with customers. [...] Pepper [...] can understand and express emotions (TechRepublic, 2016).

This description talks about Pepper in the same way we usually talk about humans. Particularly interesting for this book is the notion deployed here that Pepper was *born* and not constructed, invented, or produced. This anthropomorphic language virtually changes the status of Pepper: while being a technological artefact and therefore a "thing" in established thinking, this use of language puts it into the realm of life (born!). Furthermore, if Pepper "can understand and express emotions," then it is regarded as some higher form of life such as, perhaps, an animal, or: a quasi-human companion (Sec. 8.1).

Another example of applying anthropomorphic language and customs to the field of robots occurred in 2017. In that year, the android Sophia was granted honorary citizenship in Saudi Arabia. While this was probably a marketing gag to attract media attention and demonstrate the modernity and openness of the regime, this event raises questions about its meaning, if taken seriously. What rights would Sophia have as an honorary citizen? Even if this event seems to have been playful, it demonstrates the increasing willingness of humans to transfer human categories to autonomous technology.

While I will consider this claimed or perceived assignment of moral status (honorary citizenship) by applying specific language in-depth later (Chap. 9), only one major exemplification of this smooth shift will be mentioned here. Many publications and presentations on expected smart factories in the envisaged Industry 4.0 (Sec. 8.2.1) use the rhetoric phrase that humans and autonomous technology should cooperate "on an equal footing." The connotation is always positive: in their close and trustful cooperation "on an equal footing," a new quality of industrial production and improvement in the workplace should be reached. Taking the phrase seriously, instead of regarding it as a nice element of advertising for further digitalization, questions about its meaning arise. First, if this cooperation "on an equal footing" really did become a reality, it would be a revolution in human history: technology would reach the status and level of humans. More precisely: humans would accept autonomous technology as some new species "on an equal footing": "Robots are a new kind of entity, not quite alive and yet something more than machines" (Prescott, T., endorsement in Gunkel, 2018). The traditional master–slave or subject–object relation between humans and their technological artefacts, which involves a categorical distinction, would be transformed to a new relation "on an equal footing." In ethical respects, this would imply accepting autonomous technology as equal partners in moral discourse, e.g., for deliberating about improving processes in a smart factory. As partners in discourse, autonomous technology could claim rights, be accepted as equal, be tackled in a fair and respectful way, and so forth. In this direction, "robot rights" have been postulated (Gunkel, 2018), claiming that humans should have moral obligations toward autonomous machines, similar to human rights or animal protection rules. Robot rights could include the right to exist and perform their own missions, the right to life and liberty, freedom of thought and expression, and equality before the law, similarly to humans. Then indeed, human and autonomous robots would act "on an equal footing." However, the question remains about the meaning of this assignment and the intentions and diagnoses behind it, as well as regarding the motivation of why humans should assign dignity and moral status to fabricated things at all, and why they should accept them (Sec. 8.5).

This question leads us back to looking for the reasons how and why specific entities are subsumed as humans while others are not. We already discussed, as a thought experiment, the creation of an artificial cat with respect to its moral status (Sec. 4.3). We are used to assigning a cat the moral status of an animal and for it to be covered by animal protection rules. This is done because we know that it is a cat because it was borne by another cat. An artificial cat, however, would be manufactured in a lab, as an autonomous technology. In the case of this artificial cat having exactly the same properties and abilities as a biological cat, many people will say: "Okay, let's consider it as a cat in the familiar sense because we cannot distinguish between this artificial cat and a natural one – they have exactly the same abilities and behavior." Others might say: "This is an artefact designed by humans just like a machine, therefore let's characterize it as a mere technical object without the protection of laws on animal rights and welfare." This example allows reflection on the reasons why we apply and assign a specific moral status and value to animals and to humans.

The International Declaration of Human Rights approved by the United Nations in 1948 talks about humans, without any reference to the capabilities humans should have in order to be accepted as humans. It takes humans to mean just members of the species of humans. Thus, it refers to a biological category rather than to capabilities. This rule for assigning rights to humans only has been criticized as "speciesism," an unjustified and egoistic self-positioning of humans over other species (Singer, [1975] 1990; Ryder, 2000). However, it is this assignment which ensures that those humans with only fragile human capabilities, such as coma patients, very small children, and severely disabled persons, have the same human rights as others. This moral equality, giving the same weight to each and every human independent of their properties and capabilities, is at the core of most approaches to ethics and justice (Perelmann, 1967; Walzer, 1993; Rawls, 1971).

Transferring this model to artificial cats or robots would imply that these do not belong to the species of either cats or humans. If people postulate assigning rights to robots, they apply a completely different criterion: having particular capabilities rather than belonging to a species. Moral status, perhaps even rights, would be given to robots and other autonomous technology as soon as they became able to simulate and emulate human behavior (to a given extent). The ontological difference, however, building on biology *versus* technology (Chap. 2), remains. Therefore, the major question is whether autonomous technology gives rise to modifying or even completely converting established categories for classifying humans and other living organisms. If this conversion – e.g., regarding entities with specific capabilities such as autonomous problem-solving and learning to be human – was to happen in the future, then the question would arise of the moral status of biological humans *without* those capabilities – e.g., coma patients, as mentioned above. A corresponding question would arise about the boundaries and thresholds: what capabilities must an autonomous technology or a human being have, at minimum, in order to enjoy particular rights as a moral entity? The implications of a turnaround with respect to the criteria for assigning particular rights clearly demonstrate how far the question of assigning a moral status to autonomous technology reaches.

Summarizing, the question of a possible moral status for autonomous technology is completely different from the question of the moral status of technically enhanced or modified living organisms. In the latter case, e.g., for animals (Chap. 5), the moral status was already assigned before the technical intervention, e.g., based on pathocentric ethics. In this chapter, however, the question is about possible newly emerging needs to invent and assign an unprecedented type of moral status to technology by applying categorically different criteria for this assignment. The resulting questions, mentioned above, clearly demonstrate the emergence of heavy normative uncertainties if this pathway should be chosen in the future.

8.4   Responsibility Considerations

Responsibility analyses, reflections, and assignments should be performed with respect to the empirical, ethical, and epistemic dimension (EEE model; cp. Sec. 3.3). In cases of autonomous technology, the empirical dimension perhaps not only comprises *human* responsibility. Postulates have been expressed to consider the responsibility *of technology*, or to speak of responsibility distributed across humans and technology. After rejecting this postulate (Sec. 8.4.1), the question of human responsibility for actions taken by autonomous technology will be addressed (Sec. 8.4.2).

8.4.1   Responsibility Assignment to Autonomous Technology?

As soon as technology becomes autonomous in a certain sense, the question of responsibility assigned to technology, i.e., robots, arises. Can robots or other autonomous technologies be responsible, or made accountable in cases of failure, damage or problems caused by their activities? In order to explore possible answers a closer look at the foundations of human responsibility seems helpful.

Human autonomy is deeply interwoven with responsibility (e.g., Jonas, 1979; Gianni, 2016). Responsibility builds on autonomy, which means freedom to make a choice among alternative options in acting and decision-making (e.g., Dennett, 1984; McKenna and Perelboom, 2016). If, however, human activity was *determined* by our brains, as some brain researchers say (e.g., Haggard, 2019; Roskies, 2013; Banks and Pockett, 2007), no freedom of choice would exist. Responsibility and accountability, merit and guilt would no longer be meaningful notions for describing or orientating human practice. As a consequence, the concept of human action would become meaningless: acting would be reduced to behavior determined by physical, chemical, and physiological processes. Hence, the capability to act in combination with having a choice is a *necessary* precondition to assign responsibility. The following definition of action contrasts action with behavior (following Janich, 2001):

(1) actions can be attributed to actors (as the cause of the action) from the perspective of an observer (causality);

(2) actions can be carried out or omitted on the basis of reasons (freedom);

(3) actions can succeed or fail, i.e., there are criteria and conditions for success which are commonly explained in instrumental rationality of means and ends (rationality).

Inasmuch as humans describe themselves as being capable of acting, they regard themselves as social beings who can develop, discuss, assess, and weigh up possible actions, choose among alternative options for action, carry out the actions, and reflect about consequences and responsibility. With reference to these actions, we can speak of reasons, consequences, and responsibilities. The concept of action, as distinguished from mere behavior, is an essential element of the (modern) human self-constitution (Dennett, 1984; Habermas, 2005).

Robots, taken as an illustration of any autonomous technology, can "autonomously" cause effects in the real world, e.g., by driving an autonomous car. In order to avoid the term "action" at this point, these effects can be regarded as consequences of the robot's activities. Hence, item (1) from the list above obviously is fulfilled (actors as the cause of action). Also item (3) usually will be fulfilled for autonomous technologies, in particular while processing the cybernetic loop of adapting themselves to their environment in order to perform their tasks (success or failure) (Sec. 8.2.2). Failures or deficient consequences with regard to predetermined goals will be discovered by monitoring and feedback loops as well as evaluated in learning cycles in order to improve themselves for the next round of the loop.

Item (2), however (freedom to make choice), demarcates the crucial difference between humans and autonomous technology, according to my analysis. This difference mirrors the distinction between action and behavior, or between freedom and determinism. It would disappear if the brain researchers mentioned above were right. If humans were nothing other than programmed computers, as is formulated in machine models of humans (Mainzer, 2015), the distinction between action and behavior would no longer apply to humans.

The crucial question, therefore, is: can autonomous technology make a choice between alternatives, driven by rational reasoning? Many people would probably agree that it can. Planning robots are able to undertake several activities and can order them in different way, e.g., to deliver mail to different addressees in the most efficient and fast manner. Accordingly, they have a choice, haven't they? If yes, then robots would be able to act in a sense similarly or even equal to humans, and then there wouldn't be any reason not to assign responsibilities to them. As a consequence, robots would be made responsible in case of damage caused by them, made liable and obliged to compensate the victims, or would even be brought to court and, perhaps, jailed. The end of the argumentation sounds strange and counter-intuitive. But, where is the fallacy in the argumentative chain?

My thesis is that a robot selecting a specific activity among a set of alternative options, e.g., determining the route for delivering mail, is different from humans making a choice, even about the same activity. Assuming that humans "really" have a choice[5] the question is whether the robot "really" has a choice, too. According to the working principles of autonomous machines so far, they factually do not have a "real" choice, thereby contradicting the *prima facie*

---

[5]The question whether humans "really" have a choice and what this can mean is subject to a vast body of literature in the philosophy of mind, in the neurosciences, and in related areas.

assumption mentioned above. The reason for this statement leads back to the general understanding of what technology is. Let us assume a robot confronted with the necessity to act in a particular situation in space and time by selecting a particular action. In this situation, the current state of the robot will be given by:

- the actual status of the hardware and software of the robot, including its AI algorithms;
- a set of data collected by the sensors observing the robot's environment;
- some pre-occupations determined by humans, e.g., criteria for optimizing the route for delivering mail in the case of a messenger robot (avoid a certain route at certain times);
- perhaps some further pre-configured elements (don't go over a certain speed down the slippery hill if it is raining).

Due to my understanding, the result of the selection process *will be determined* by the given constellation. If the output of the selection process would, on the contrary, *not* be determined by the respective input data, then the autonomous robot acts as a kind of random machine. If it arrived at different outcomes with identical input in all of the categories mentioned, the deviation would be neither understandable nor desired. It would not be understandable because no reason could be found for the deviation – this is exactly what random means. And it would not be desired because an autonomous technology which produced random output would not be able to fulfill the tasks it should serve according to its mission. Fulfilling tasks in an expectable way excludes the possibility that the computer "does something arbitrarily on its own." As long as the robot works on behalf of humans and according to human purposes, the requirement is that its outcomes are determined by its actual state and the input data collected.

According to many interpretations from philosophy (e.g., Hubig, 2007) and sociology, the social function of technology is to stabilize expectations and thereby to stabilize fields of human practice. Realizing this feature necessarily prevents robots from doing unexpected and random things on their own. Deviations from desired or expected behavior would simply be regarded as a technical failure.

This could be regarded as a rather boring result, and in a sense it is. But what we can learn from it is that autonomous technology acting on its own behalf or as a random machine could not fulfill human expectations with regard to technology. But perhaps there is some uncharted territory behind this position. A robot doing things on its own could perhaps be regarded as something different from technology – a new form of life (cp. Chap. 9). Researchers could investigate the behavior of this robot similarly to investigating the properties of a new species found in the Amazon, which would also show unexpected behavior. In this way, the formerly strict boundary between living organisms with their own behavior and technologies dedicated to predictably fulfilling human purposes would become fluid. We will consider this really interesting case later (Chap. 9).

At this point of our reasoning, however, we go back to the title question of this section, which is about assigning responsibility to autonomous technology. So far, we have recognized that the "autonomy" of robots is different from human autonomy, with the major difference concerning the capacity to make "real" choices. If this is correct, robot calculations must be different from human reasoning. But what exactly is the difference? This question refers back to the question of the nature of human reasoning. Models aiming at understanding usually try to understand functional principles analogously to a computer: the brain is modeled as an organ which collects and evaluates data coming from its sensors, e.g., its eyes and ears; stores and evaluates data; draws conclusions; and determines actions to be taken. Modeling the brain in this way opens up certain ways of understanding and for therapy in case of malfunctions (e.g., Glees, 2005; Kolb and Whishaw, 2013). However, modeling remains modeling and can never fully grasp the modeled object (Manzano, 1999). We have to make a distinction between the model and the modeled object (Janich, 2002; Nick *et al.* 2019). Knowledge drawn from models and modeling must not be confused with direct knowledge about the modeled object. There will be overlap but possibly also difference, depending on the case.

Ignoring this difference and considering the knowledge gained by modeling the brain as a data processing machine as immediate and direct knowledge about the brain as modeled object, would imply identifying the model with the modeled object. In this fallacy, the brain is regarded as nothing more than a data collecting and evaluating machine doing (mostly statistical) calculations. Then, the above-mentioned thesis applies, that its output (actions to be taken) should be determined by the input data collected by sensors and the actual status of its "algorithms." Accordingly, human autonomy would be no different from the "autonomy" of robots, and there wouldn't be any reason to assign responsibility to humans but not to robots. However, I insist that this is a fallacy due to misunderstanding what a model is and what it can provide.

Readers might have the question in mind: What is the brain "really," beyond the models that brain researchers and neuroscientists have been developing? Responding to this extremely complex question is far beyond the scope and intention of this book, but it is interesting to note that the question of responsibility ends at such a fundamental point: If humans function similarly to computers, then action and decision-making will be either determined or random. Then, talking about responsibilities would no longer make any sense. Reflecting and assigning responsibilities necessarily needs to share the presupposition that human reasoning and action are neither determined by input data nor random but *oriented to reasons* on behalf of the humans themselves. Choosing among alternatives must not be a random choice but must reflect reasons, arguments, and criteria, which partly will only be developed and weighed up during the reasoning process. This is the core dichotomy between naturalist and culturalist understandings of the human brain (Hartmann, 1996) and of the categorical difference between calculation and reasoning (McKenna and Perelboom, 2016; Habermas, 2005).

This dichotomy can easily be illustrated by the following thought experiment (Grunwald, 2016c) closely related to Searle's "Chinese room" argument (Searle, 1980). First, a human being is observed while performing some action. Then, this human being is replaced by a robot acting in a functionally equivalent manner. If the external observer now considered what the human and then the robot does, the result of their interpretation should be the same for the robot as for the human being: both are regarded as intervening into their environment, thereby fulfilling items 1 and 3 above. However, in spite of the identical steps at the phenomenological level, the external observer will note that a crucial difference remains. In the thought experiment, neither the human being nor the robot observed was asked about their/its activities, tasks, diagnoses, and reasons. The interpretation was made solely from the observer's external and phenomenological perspective. This seems to be artificial concerning human beings: why not ask acting persons about the reasons for their choices, about possible alternative options, for the criteria of assessment, and so forth, thereby accepting them as participants and actors? In the case of the robot, however, asking for its reasons does not make sense. In this case, the programmer should be asked for the rationale behind the algorithms the robot is based upon and about the calculations it was expected to perform. This observation directly leads to the "Chinese room" argument developed by John Searle: "The appropriately programmed computer with the right inputs and outputs would thereby have a mind in exactly the same sense human beings have minds" (Searle, 1980). This observation illustrates a deep-seated difference between humans and robots, even in a case where both are doing the same actions, in a phenomenological sense: the robot does not own the perspective of a *participant*.

This difference will become even clearer by considering a messenger robot as a specific case. A robot which is functionally equivalent to the human messenger, i.e., it brings the same messenger performance, plans the errands and the solution to problems which arise in a specific sense and under predetermined initial conditions. The human messenger plans in a similar manner, and with probably similar criteria and perhaps similar outcomes. So far, this example is identical to the story of the previous paragraph. But now we consider an unexpected situation. A human person lies on the floor and constitutes an obstacle to the messenger continuing to do their job. Currently, the robot would carefully circumvent the human obstacle and continue delivering mail. The human messenger, however, will wonder why the person is lying there. They would immediately stop doing their job and instead help the person in an appropriate manner. While the messenger robot is committed to its role as a messenger through programming and control architecture, human messengers can abandon this role for what they perceive to be good reasons. The ability to abandon the normal plan and switch to a completely different track of action for reasons occurring during operation proves to be central to the distinction between humans and robots. Robots can be prepared for a set of foreseeable unusual situations but not for the entirety of all of the unusual situations that occur in daily and often chaotic life. Therefore, a considerable asymmetry remains between the messenger robot and the human messenger, demonstrating that doing calculations within a programmed framework is different from human reasoning and action.

Accordingly, the conclusion of this section is that assigning responsibility to autonomous technology does not make sense, insofar as robots or other autonomous technologies do not develop their own will, which would allow them to break out of the limited world of their statistical calculations to do something "on their own." Thereby, I subscribe to the preceding philosophical distinctions made between activities driven either by causes or by reasons (Habermas, 2005), with humans seen on the side of reasoning. Therefore, the realm of responsibility remains with humans only. This holds at least as far as autonomous machines do not develop any consciousness, intentions, or a kind of will which allows them to make choices for reasons, rather than being determined by data.

However, one restriction remains. There is no proof available that humans "really" are categorically different from calculating and data processing machines. The word "really" sounds nice, but its meaning has been extremely contested in the history of philosophy. According to philosopher Immanuel Kant, there is no possibility at all to make sense of the word because we do not have access to the "real world" as such. Therefore, my argumentation corresponds to an "as if" mode of operation: I talk about responsibilities, as I did in the preceding chapters, "as if" humans were free (to a certain extent) to make choices according to reasons and arguments, while I do not see this situation as given for autonomous technology. Therefore, with respect to actions, e.g., for autonomous vehicles in road traffic, responsibility must ultimately be assigned to humans rather than to the machines (Wallach and Allen, 2009). This result raises the question of who is, or who should be, made responsible for the behavior of autonomous vehicles in human/robot interactions, in automated road traffic, or in the smart factories of Industry 4.0.

8.4.2 Human Responsibility for Autonomous Technology

If responsibility assignments with respect to interventions of autonomous machines into the real world remain with humans, the main questions are about the addressees of those assignments, the distribution of responsibility, the criteria for assigning responsibility in a specific manner, and so forth. This is a highly complex field in each case, which requires contextual adaptation to the respective situation and configuration. Therefore, I can only develop some more general considerations in this section.

According to the framing of responsibility applied in this book (Sec. 3.3), assignments are elements of social practice. Responsibility assignments should contribute to "good" social practice in the respective field, with a strongly context-dependent meaning of "good." Obviously, the advent of autonomous machines gives rise to normative uncertainties and corresponding challenges for an adequate responsibility regime. The rationale of reflecting responsibility then is to overcome unclear responsibility distributions in order to reestablish "good" practice in the affected field. The search for new responsibility regimes ultimately serves a practical purpose: ensuring good practice in newly emerging human–machine configurations, according to the familiar sets of criteria such as human rights, trust and trustworthiness, issues of democracy and individuality, sustainable development, and so forth.

The introduction of automated and autonomous vehicles, as an example, raises the question, amongst others, of who shoulders responsibility in the event of an accident (Hevelke and Nida-Rümelin, 2015). For practical reasons, the question of who is legally or morally responsible has become urgent over recent years (Maurer *et al.*, 2016). Autonomous vehicles will not be established in practice before this question has been answered unambiguously and in a manner that will stand up in court. Determining an "adequate" responsibility distribution as well as deriving corresponding liability regulation is essential, whether in driving, in caregiving, in delivering mail, or in industrial production.

The German system of liability currently assigns the risk of a road traffic accident in the final instance to the keeper or driver of the vehicle (following EK, 2017). The manufacturers of the vehicle are liable within the scope of statutory product liability. Highly automated and autonomous vehicles, however, are also subject to more far-reaching determinants and hence need a more complex responsibility regime. Not only the keepers and manufacturers of the vehicles but also the corresponding manufacturers and operators of the vehicles' assistance and autonomous technologies have to be included in a regime of responsibility and liability sharing. A new distribution of the responsibilities shared by manufacturers, suppliers, and operators of components, of the programmers and engineers, of managers, regulators, and ethics commissions involved is required:

> In the case of automated and connected driving systems, the accountability that was previously the sole preserve of the individual shifts from the motorist to the manufacturers and operators of the technological systems and to the bodies responsible for taking infrastructure, policy and legal decisions. Statutory liability regimes and their fleshing out in the everyday decisions taken by the courts must sufficiently reflect this transition (EK, 2017, rule 10).

Together with the shift of responsibility away from the traditional system focused on drivers and keepers of cars, to the parties responsible for the technological system within the context of, e.g., product liability, a further challenge arises. In cases of highly automated cars being able to drive autonomously, but which also can be driven by human drivers, we have to distinguish between two modes of operation: (a) the human driver has full control and is responsible in the traditional manner; (b) the computer has full control, governed by a new system of responsibility and accountability. In this system, handovers of control are possible. The human–machine interface must be designed such that it is clear who is driving the vehicle at any point in time, in particular while transferring control:

> In the case of non-driverless systems, the human–machine interface must be designed such that at any time it is clearly regulated and apparent on which side the individual responsibilities lie, especially the responsibility for control. The distribution of responsibilities (and thus of accountability), for instance with regard to the time and access arrangements, should be documented and stored. This applies especially to the human-to-technology handover procedures (EK, 2017, rule 16).

Hence, we have to distinguish between two completely different responsibility regimes, with both of them obliged to meet the same rationale, of (1) ensuring safe and reliable transportation without endangering others, and (2) taking care in case of technical malfunction. While modern societies have developed and optimized the traditional system over more than a century, the new regime is still under construction. In this regime, the manufacturers will be responsible for the functional safety of the systems, based on certain datasets. The responsibility for content and quality of safety-related data, which are exchanged with the vehicle via the traffic-based information and communications infrastructure must be guaranteed by others who are accountable for the data infrastructure of the system (EK, 2017, 27: diagram). Determining safety and security standards for the mobility services could be realized by introducing certificates supervised and controlled by a public agency.

Major attention must be paid to monitoring problematic situations and to learning from them, according to the cybernetic learning cycle mentioned above. To prevent errors and to ensure the safety of all road users, there should be a continuous analysis of hazardous situations relevant to misperception and problematic behavior by the vehicle (EK, 2017). A public institution could be made responsible for supervising the results and drawing conclusions.

> [...] it would be desirable for an independent public sector agency (for instance a Federal Bureau for the Investigation of Accidents Involving Automated Transport Systems or a Federal Office for Safety in Automated and Connected Transport) to systematically process the lessons learned (EK, 2017, rule 8).

Responsibility is not restricted to the realm of engineering, manufacturing, managing, regulating, controlling, and monitoring. Because autonomous vehicles are, as an element of road traffic, an important part of public life, the general public also has to be not only informed but should, depending on the situation, also be involved, in the sense of an inclusive technology assessment (Grunwald, 2019a, Chap. 4).

> The public is entitled to be informed about new technologies and their deployment in a sufficiently differentiated manner. For the practical implementation of the principles developed here, guidance

for the deployment and programming of automated vehicles should be derived in a form that is as transparent as possible, communicated in public and reviewed by a professionally suitable independent body (EK, 2017, rule 12).

This brief consideration results in an expansion of the range and distribution of responsibilities. Even if autonomous technologies cannot assume responsibility, either today or in the foreseeable future, the complexity of responsibility assignments still strongly increases. The best example is car-driving, comparing the traditional driver-/keeper-dominated assignment compared to new responsibility regimes including several groups of actors with shared responsibilities. But, also in smart factories of the Industry 4.0 vision, responsibility will be distributed among many more actors in the background than in today's factories.

A major challenge involved with this strongly widened range of shared responsibilities is to organize a transparent responsibility regime with a clear distribution. The negotiation and determination of the new and "adequate" distribution of responsibilities as well as accountabilities and liabilities in a unanimous and clear manner is essential for ensuring "good" social practice. Arranging this is a highly interdisciplinary task involving engineers, computer scientists, legal scientists, ethicists, experts on the governance and organization of complex socio-technical systems, managers, and so forth. This interdisciplinary reasoning must be embedded in a transdisciplinary environment, consisting of regulators, democratic institutions, civic society, worker unions, stakeholders, and even citizens, depending on the context.

8.5 Autonomous Technology: Loss of Human Autonomy?

The public and ethical debate on the subject of this chapter – increasingly autonomous technology in the midst of society – is accompanied by skeptical voices and concerns of a loss of human control, a loss of human autonomy, and of the increasing subordination of humans under technology. This was discussed as an issue at the occasion of self-driving cars with human drivers controlled by the onboard computer, which would permit humans to take over, or not. Even if this is no more than a specific application case, it illustrates a much broader issue. Does the increasing autonomy of technology factually or necessarily decrease or restrict human autonomy? By applying a cake model of autonomy, this effect sounds plausible: giving more of the cake to technology would diminish the part of the cake remaining for humans. However, things are more complex, and we have to look behind the *prima facie* image. In the following, the development of human autonomy in the context of autonomous technology will be explored, as a loose collection of heterogeneous aspects, some of which might come as a surprise.

(1) *Loss of human autonomy through loss of transparency*

An important but largely overlooked threat to human autonomy results as a side effect of digitalization enabling autonomous technology. The increasing complexity of Big Data, AI algorithms, huge models, learning systems, and self-modifying software restrict the traceability and understandability of the behavior of autonomous systems. Many observers, as well as computer scientists, warn of a loss of oversight and control over autonomous systems. Their concern is that it may become increasingly difficult, and perhaps even impossible, to trace and scrutinize how the *results* of complex operations, e.g., AI algorithms which mine Big Data, correlate with the input data, the state of the algorithm, and the models used. Those systems, therefore, can develop into "black boxes" with a decreasing capability for humans, even computer scientists, to understand the internal functions of the "box." In such a situation, transparency gets lost. If the outputs of the autonomous system can no longer be transparently related to the input data, then, at the end of the day, only *belief* in the system would remain. If it became impossible to trace how the results emerged, then we would only be left with the choice to either believe in the results, or not. This could herald a return to pre-modern times, with black box AI systems acting as the new oracles, without transparency and understandability. Loss of those crucial gains of human development, e.g., pushed by European Enlightenment, would indeed threaten human autonomy.

In a situation where there was no opportunity to distinguish whether the system produced excellent results or nonsense, this could easily lead to adverse scenarios. Decisions based on nonsense, e.g., mere artificial correlations (Vigen, 2015), could be disastrous in many fields, including medicine, politics, and the economy, as well as for self-driving cars. This is not only a dystopian possibility for the future. Long before the complexity of today's autonomous technology was reached, an effect of this kind caused heavy damage: among the origins of the 2008 World Bank Crisis, with the subsequent world recession, was too much trust placed in huge models to calculate the risks of financial transactions. Bankers responsible for checking the risks of derivate trade did not understand, but simply believed in, the outcomes of the calculations, which showed themselves to be false (O'Neil, 2016). This story clearly warns us against placing blind trust in those systems. Blind trust, in this respect, would constitute voluntary abandonment of human autonomy, without any necessity to do so.

(2) *Loss of autonomy; opening up new autonomy?*

In many cases, fear of the loss of human autonomy uncovers a contingent and biased perspective on autonomy itself. The case of the subordination of a human driver under the supervision of an autonomous car serves as an example (Sec. 8.3.2). Many people feel autonomous while controlling their car themselves. Giving control to a computer causes pain to them because they feel they have lost this freedom and autonomy.

However, the autonomy of humans while driving their cars is a special case. They can indeed control all the cockpit functions, regulate speed and thereby exert power, and demonstrate this to others. But the reality often shows us a different picture. Most road traffic is gathered on crowded streets, where signals strictly regulate the traffic. The driver always has to carefully observe the environment of their own car, the behavior of other participants in the traffic, the weather, and the status of the road surface. All the time the driver must take decisions in order to enable a safe trip, in particular under adverse conditions of weather, street, and traffic. From this perspective, the autonomy of the driver is strongly limited because they are part of a complex system, which requires strict observance of its rules. The human driver mostly does not act truly autonomously but has to *react* and is a kind of slave of the respective status of the traffic situation and system.

Humans in self-driving cars present a different picture. Human autonomy now consists of being free from the continuous need to keep close attention on the respective traffic situation and to constantly make the necessary decisions. They are free to use this time for other purposes while being driven by the car. They can autonomously decide what they want to do: read a book, watch a movie, join a virtual business meeting, or simply sleep. From this perspective, self-driving cars increase human autonomy rather than restricting it.

This example demonstrates that the frequently feared subordination of humans under autonomous technology should be carefully scrutinized and reflected. In some cases, widening the perspectives can, as seen above, allow for a re-interpretation. It could, for example, provide the insight that the traditional understanding of human autonomy (e.g., to drive a car) could be contingent due to a simple accommodation to cars over many decades. In this case, as was shown, the diagnosis of a loss of human autonomy could be transformed into a gain. In other cases, however, things could look differently, for example, if supremacy over inhabitants of a care home was given to care robots. It depends on the case.

(3) *Loss of human autonomy through disappearing responsibility*

Facing the complexity of distributing responsibility among several groups and actors, society could run the risk of "thinning" responsibility, or even making it invisible. Responsibility could get lost in an anonymous space of diverse actors through simple diffusion, or it could even disappear in complexity.

This issue is well known from complex responsibilities distributed throughout society (Beck, 1988) and in large institutions and companies. Over and over again, unclear, diffuse or thinned responsibilities and accountabilities lead to institutional failure. This problem will be strongly amplified in systems including autonomous technology, in particular in relation to the issue of lack of transparency mentioned above. In such systems the risk increases that human responsibility could be implicitly eradicated by trying to transfer it to diffuse technological or socio-technical systems. However, this transfer would be a mere illusion, or even a self-betrayal, because technology cannot assume responsibility (see above). Instead, it would make the human responsibility *behind* those technologies invisible: This transfer would give managers and company owners, shareholders, engineers, and computer scientists a free hand to follow their own interests without being held responsible. Loss of human autonomy in this case means leaving technology development, diffusion, and implementation to an assumed and blind self-dynamics. Agency of humans would implicitly disappear along with responsibility. The impression of an inherent dynamics of the use and spread of autonomous technology could emerge, followed by fatalist adaptation. This issue leads directly to technology determinism, as below.

(4) *Autonomy: shaping technology instead of adapting to it*

Human autonomy would come under pressure if autonomous technologies and their development over time forced humans to adapt. In current public debate, the dominant perception is that digitalization develops according to its own dynamics. Especially business representatives and politicians like to talk about digitalization as an inevitable natural phenomenon, like a tsunami or an earthquake (Grunwald, 2019b). If this were the case, society and individuals could only adapt to it, and actively shaping the emergence and enculturation of autonomous technology would not be an option. Obviously, the force to adapt to new and increasingly autonomous technology would decrease human autonomy, understood as being able to make choices.

Fortunately, this "technology determinism" (Grunwald, 2019a) contradicts simple facts. Each line of a source code for software has to be written by humans. Software runs on hardware that is also produced by humans, or by machines that have been developed and programmed by humans for this purpose. Algorithms, robots, digital services, business models for digital platforms, or applications for service robots are invented, designed, manufactured, and deployed by humans, who usually work in companies, authorities, or government services. They pursue certain values, have opinions and interests, and follow corporate strategy, political requirements, military considerations, etc. when developing autonomous and other technology. Others, however, have to adapt. Hence, the question is about who is on the side of the influencers and who simply has to adapt their behavior to the results of decisions made by others. Finally, this is an issue of the *distribution of power*: some decide while others have to adapt. The loss of autonomy of many who are forced into adaptation is accompanied by increased power for a few. Therefore, the feared self-subordination of humans is not a subordination under autonomous *technology* but under the *humans and institutions* making use of those technologies!

In technology determinism, possible alternatives for shaping autonomous technologies, e.g., according to ethical or social considerations, are ignored, forgotten, or even suppressed. However, this is not the *only* way of developing and using autonomous technologies in the future. Instead, the future is a space of possibilities full of alternatives. Which of them will become real depends on many decisions at very different levels, in companies and data corporations, in politics and regulation, by computer scientists and engineers, in science policy, and in the agenda-setting of research institutes. Ensuring or reestablishing human autonomy in this respect means engaging many more people and actors with their own perspectives, hopes and fears into the agenda-setting and development process. Approaches such as technology assessment (Grunwald, 2019a), responsible research and innovation (Owen *et al.*, 2013), and value sensitive design (Friedman *et al.*, 2006) are available for supporting shaping technology and innovation instead of adapting to them.

At the close of this chapter, I would like to recall some major insights achieved: (1) human autonomy and the autonomy of technology are different by category (Sec. 8.4). Technological autonomy is determined, attributed, and constrained by humans, according to purposes determined by humans. The expectations of humans as to what care robots, social bots on the Internet, or self-driving cars should provide guide their shape and the properties of the algorithms used. This also applies to AI systems used for practical purposes. Autonomous technology remains in the realm of technology. There is nothing special about autonomous technology so far. (2) However, the assignment of responsibility in socio-technical configurations involving autonomous technology becomes much more complicated because of its distribution across more actors. But, also in this respect there is nothing categorically new about this situation. (3) Things would only become completely different if autonomous technology developed its own intentions and could act beyond human expectations, out of its own diagnoses and objectives, and so forth. This, however, is a merely speculative story which is not valid today, nor will it be in the years, and probably decades, to come. (4) Loss of human autonomy is an interesting issue. In-depth considerations demonstrate that its loss can sometimes be considered as a gain from another perspective, while in other cases it simply depends on we humans whether or not we tolerate a loss of our autonomy, e.g., because of a lack of transparency, or due to mere adaptation.

Chapter 9

On the Future of Life

The rapid advance of science and technology creates new crossroads between life and technology and transgresses formerly strict boundaries. Human and animal enhancements, synthetic biology and artificial intelligence are all tied to the ongoing dissolution of traditional boundaries between living beings, including humans, and technology. In the case studies presented in the previous chapters, we witnessed these ongoing processes and their ethical implications and the responsibility challenges involved at different levels regarding living entities. In this final chapter, I will identify overarching observations, trends, challenges, and perspectives. In the short subchapters, I will focus on the major generalizable issues, which lie at the crossroads between life and technology.

While the story of life on Earth will continue in the Anthropocene (Crutzen, 2006), life will be increasingly affected by human influence. In addition to natural evolution, which has shaped the development of life since its beginning, humans now increasingly contribute to that evolution through technology and techno-induced impact. The deep transformation of almost all the relations between nature, humans, and technology demands understanding and reflection in order to shape that transformation responsibly. From this perspective, the future of life on planet Earth will be, metaphorically speaking, a co-creation by nature and humans, which gives an impression of the mounting challenges ahead.

Living Technology: Philosophy and Ethics at the Crossroads Between Life and Technology
Armin Grunwald
Copyright © 2021 Jenny Stanford Publishing Pte. Ltd.
ISBN 978-981-4877-70-1 (Hardcover), 978-1-003-14711-4 (eBook)
www.jennystanford.com

The discussion below will contribute to clarifying the major questions and challenges and provide some insights and orientation regarding the next steps. By heading the subchapters mostly as questions, I dedicate my reflections as steps on the long path of clarification and new orientation that lies ahead of us. The thematic consideration begins with interpreting ongoing processes, shifts, and transformations at the philosophical level (Sec. 9.1–9.4), and progresses to reflections on ethical issues such as moral status and responsibilities (Sec. 9.5). This leads to the final step to be taken in this book: presenting a proposal for enhancing transparency in the debates on the future of life, including humans and human responsibility, in the form of a "ladder of life" (Sec. 9.6).

9.1 The Ultimate Triumph of the Baconian *Homo Faber*?

An observation common to all the cases discussed in this book is the emergence of new crossroads between life and technology as a result of science-based technical intervention. Humans acting as engineers no longer address only things such as bridges, washing machines, and cars; they also engineer cells, genomes, and entire organisms, perhaps even evolution itself. The notion of "engineering life" corresponds to the philosophical background of the well-known Baconian approach (Schäfer, 1993), which can be summarized in the imperative to establish human dominion over nature (Grunwald, 2016a). In this approach, the advance of humanity is seen as completely in line with the mainstream of the European Enlightenment, aiming to realize more and more human autonomy and independence from any restrictions of nature, natural evolution, tradition, or religion. Humans should empower and enable themselves to shape their environment and their living conditions to the maximum extent possible, according to their values, preferences, and interests. Here, the main concept of what humans are supposed to be is the *homo faber*, who intervenes into the world through self-invented technology. The approach of humans *as engineers* who manipulate, shape, and create life may be regarded as a notion which adds to the *homo faber* model by involving science, in particular technoscience (e.g., Asdal *et al.*, 2007), and extending this to the *homo creator* model discussed in synthetic biology (Boldt and Müller, 2008).

The ultimate triumph of *homo faber*, who sets out to manipulate the world atom by atom, according to human purposes, was expressed in the early debate on nanotechnology (NNI, 1999). In a revival and a radicalization of the physical, even mechanistic, reductionism of the nineteenth century, this new *homo faber* is expected to bring everything under control, including the realms of life and society. This picture applies to synthetic biology (Chap. 4), to the ongoing modifications of animals according to human purposes (Chap. 5), to engineering the human genome (Chap. 6), to far-ranging human enhancement stories (Chap. 7), as well as to the cosmos of digitalization and artificial intelligence (Chap. 8). The *homo faber* story, or the more recent notion of *humans engineering the world*, e.g., in the phrases of geo- and climate engineering, seem to be common to visions regarding new crossroads between life and technology. This story, which fits well with the diagnosis of the Anthropocene (Crutzen, 2006), has been interpreted as an underlying metaphysical program:

> The aim of this metaphysical program is to turn man into a demiurge or, scarcely more modestly, the "engineer of evolutionary processes." [...] This puts him in the position of being the divine maker of the world [...] (Dupuy, 2005).

Several visionary, techno-optimistic and sometimes even utopian authors expect solutions to all of the world's problems of health, environment, and climate, including questions of poverty, hunger, justice, crime, and war, to emerge from further scientific and technological progress in this Baconian sense. In this regard, the immense progress made in science and technology as a result of human creativity and productivity could give us reasons to be proud and self-confident. Analogous to God in the Genesis story in the Old Testament, humans, as the creators of technology, could look back at their work at the end of the day and say that it was good.

But something sounds wrong with this story. Questions remain to darken the bright story of progress, and even salvation, through technology. Increasing dominion over nature does not lead to full control. The Covid-19 virus, which has caused the worldwide Corona pandemic of 2020, clearly indicates the limits of human control. In a mixture of cautious, helpless, responsible, and panic reactions humans tried to stabilize the situation as the virus spread, facing huge uncertainties under the pressure of natural processes. The main reaction worldwide was social distancing, which is really not a high-tech, or even new, strategy, as it has been used by humans for millennia in the face of infectious disease. In fact, high-tech measures were not of much utility, as while the millions of low-tech masks required were not available in many industrialized countries, high-tech civilization acted as an accelerator of the pandemic: worldwide logistics and mobility enabled the rapid spread of the virus. Modern technology is therefore not (only) an expression of human dominion over nature but also increases human vulnerability to unexpected events, as well as dependence on the trouble-free functioning of the technological environments that humans have become used to living in. In this respect, increasing human vulnerability, e.g., with respect to blackouts of the electricity supply, or the Internet, or of maintaining a reliable food supply, is a twin of the dominion over nature gained by technology. Dependence on nature seems to be replaced by the dependence on technology. However, the independence from nature gained is accompanied by new dependencies on nature at other levels, indicated by, e.g., the impact of climate change, which creates and fuels awareness of human dependence on friendly climate and environmental conditions.

Serious problems have been well known for decades: pollution, climate change, loss of biodiversity, scarcity of resources, and many other unintended side effects of the techno-based economy and civilization (Grunwald, 2019a). Many people are plagued by doubts about what the techno-scientific advance that humans have created means to us and for the future of life on Earth. Concerns have arisen that we could be the victims of our own success in the medium or long term. The diagnosis of an ultimate triumph of the Baconian *homo faber* is accompanied by gnawing self-doubt, which prevents many from looking back favorably at human achievements at the end of the day and feeling satisfied. Deep ambivalence remains concerning the status reached so far and the pathway ahead of us. Concerns about the severe challenges and problems resulting from human remodeling of the entire planet escalate to the unintended self-destruction of humankind (e.g., Jonas, 1979). While welcoming the many opportunities and benefits provided by new developments at the crossroads between life and technology, some modesty could and even should be the counterpart of enthusiasm. Modesty includes, in particular, paying attention to possible unintended effects, avoiding naïve techno-optimism, and addressing responsibilities at many levels, as discussed in this book.

9.2   The End of Nature (1): Human Impact on the Future of Life

A millennia-old perception of the ontological order of being has dug itself deep into humans' cultural self-image, at least in those parts of the world influenced by Western thought. In the Genesis story in the Bible, for example, the ontological order of the world described in its first chapter is grounded in the sequence of acts of creation by God, at the end of which stands the creation of humanity. When, at the beginning of the twentieth century, Gustav Mahler starts his third symphony with impressions from the rocks and proceeds via plants, animals, humans, and angels to finally reach love, this portrays the Old Testament order in pantheistic vestment, no longer having in mind the singular act of creation but maintaining the same ontological order. This order of being was and still is often interwoven with hierarchical value judgments, which assign different values to observable entities such as plants, animals, and humans. In spite of being deeply rooted in culture, this order is challenged by the recent steps to making traditional boundaries between life and technology more permeable. We can see this at the occasion of synthetic biology, of technologies such as CRISPR-Cas9, which can be applied to plants, animals, and humans, and of transgressing boundaries between technology and life through digital technologies.

The ontological separation of all entities on planet Earth into living and non-living objects or, in an even more ambitious notion, into animated and unanimated objects is usually traced back to Greek philosopher Aristotle, who introduced a fundamental distinction between the natural (the realm of the grown) and the cultural (the realm of the human-made) about 2500 years ago. While non-living entities were regarded as merely subject to external influence and physical and chemical laws, an intrinsic source of growth and movement was assigned to living entities. "Living beings are defined as natural objects which have the source of their growth and movement in themselves" (Funk *et al.*, 2019, 177).[1] The position of vitalism corresponds to this consideration of life as something ontologically specific: "living organisms are fundamentally different from non-living entities because they contain some non-physical element or are governed by different principles than are inanimate things" (Bechtel *et al.*, 1998; cp. "Vitalism." *Wikipedia.* https://en.wikipedia.org/wiki/Vitalism; accessed 2 June 2020). Technology, in contrast, is regarded as belonging to the realm of culture because it is human-made, rather than grown "by itself." In spite of the fact that philosophical vitalism today is regarded as no longer tenable (e.g., Caplan, 2010; Gutmann and Knifka, 2015; Gutmann, 2017), the ontological separation of the world into living and non-living entities with no bridge between them is still deeply embedded, at least in Western culture.[2]

In this worldview, life belongs to nature, in particular through the assumed inherent drive regarded as the cause of growth and movement of living entities, which is absent in non-living ones. At this point, however, a deep-ranging challenge emerges when considering the case studies presented in this book, where we have identified the overarching movement of technology into the realm of naturally living entities. Referring to Aristotle's distinction between the natural (the realm of the grown) and the cultural (the realm of the human-made) implies that culture (in the form of human-made technology) is colonizing nature. Nature becomes "occupied" by humans. The purely grown is no longer available, while the entire planet demonstrates human impact – which is precisely the observation motivating the diagnosis of a new era called the Anthropocene (Crutzen, 2006). Human influence is detectable even in landscapes, oceans, animals, plants, bacteria and viruses far from any human civilization and influences the chemical composition of environmental compounds, living entities, ecosystems, and the climate. Consequently, the Aristotelian distinction breaks down in

---

[1]This quote and some other quotes referred to in this chapter have already been mentioned in the previous chapters. They are reprinted here in full text in order to improve readability and usability of the chapter.

[2]Accordingly, life is frequently assigned an intrinsic value or even some dignity, which is not assigned to, e.g., crystals and rocks, for example, in the widespread philosophy of Albert Schweitzer.

its either/or dichotomy: nature in this sense of including natural life disappears, while its relics become part of human culture because they are carrying traces of human impact.[3]

Concerning the future of life on Earth, human impact is, on the one side, *constructive*: humans create modified or even new forms of life, which is the main area of interest of this book. Synthetic biology, genetic engineering, enhancement and artificial forms of life in the digital arena are examples of humans entering the formerly "natural" world through their own intentions and realizations. Thereby, humans add to ongoing natural evolution. The "made" enters and enriches the "grown."

However, the dark side of the coin must not be ignored. There is a highly *destructive* side of human impact on the presence and future of life on planet Earth. The extinction of huge numbers of species, the destruction of ecosystems such as tropical rain forests, the devastation of large areas for mining and industry, the monocultures in industrialized agriculture which threaten biodiversity, the suffering of animals exploited for mass production of meat, the persistent problems caused by uneven human development in many regions of the world, and so forth. All of these developments darken the future of life on Earth and reduce diversity, often in an irreversible manner.

The end of nature in the Aristotelian sense is, therefore, ambivalent. While some visioneers in Baconian optimism are dreaming of an evolution 2.0 under full human control, the unintended consequences of the technological advance and of the use of its outcomes in the economy are obvious. Taking this ambivalence seriously, there is no way to naively applaud a presumed jump from (natural) evolution 1.0 to (human-made) evolution 2.0. Instead, observing the diagnosis of the end of nature in the Aristotelian sense, we might speak of a co-evolution of natural evolution and human impact with, as is an implication of evolutionary models in general, an open-ended future.

---

[3]Romantic ideas of wilderness and pure nature, of the beauty of untouched nature, and desires of a return to nature seem to be more a reflex reaction to the loss of nature in the Aristotelian sense than something which could seriously happen. While romantic feeling will still have a place in human culture and perspectives on the world, the time is ripe to accept the end of nature in its traditional understanding. Instead of nostalgic feelings while looking back to presumed good old times, the challenge now is to look to the future without losing the legacies of the past.

9.3 The End of Nature (2): Dominance of the Techno-Perspective

There is a second observation indicating and supporting the thesis of the end of nature. We can identify it by looking at what transgressing the boundary between life and technology means. Principally, the transgression can happen in two directions: (1) technology increasingly follows the models provided by natural life and adapts to it; (2) natural life comes under the realm of technology. I will demonstrate that option (2) is the dominant move contributing to "the end of nature."

Some voices are optimistic with respect to option (1). In particular, synthetic biologists refer, more or less explicitly, to natural forms of life as a model for researching and developing new technology, for new interventions into existing forms of life, and even for new forms of life. Biomimetics and nanobionics seek to explore and understand existing forms of life and to rebuild them with technology, or to reuse the knowledge gained to develop new configurations of life and technology (Sec. 4.5). Some authors combine this approach with expectations concerning the reconciliation of technology and nature (von Gleich *et al.*, 2007), assuming that on this track a technology will be achievable that is more natural, or better adapted to nature, than traditional technology (Grunwald, 2016a). In grounding such expectations, advocates refer to the problem-solving properties of natural living systems (Wagner, 2005, 39). Philosophically, this idea goes back to German philosopher Ernst Bloch, who postulated moving toward an "alliance technology" (*Allianztechnik*), in order to reconcile nature and technology and approach peaceful co-existence between humans and the natural environment; this is the opposite position with respect to the Baconian approach of human dominion over nature (see above).

This view, however, misses a crucial point. Synthetic biology, animal and human enhancement, and genome editing are all linked epistemologically to a technical view of the world including life and are aimed at technical intervention into that world. Living systems and parts of them such as subcellular units, cells, viruses, the human genome, the nervous system, and the human brain are investigated by biology, the life sciences, and brain research in the *relationship of their technical functioning* and are thus interpreted as *technical systems* (Grunwald, 2016a). The engineering approach carries technical ideas into the natural world, modulates elements of "natural" nature such as cells, animals, and humans in a techno-morph manner and gains specific knowledge through applying this perspective (Gutmann and Knifka, 2015). Living entities are increasingly described in technological or techno-morph terms. From this perspective, they are interpreted and modelled as *machines* composed of modular components (Sec. 2.3). In the final consequence, nature is seen as machinery, both in its individual components and as a whole:

> This is where a natural scientific reductionist view of the world is linked to a mechanistic technical one, according to which nature is consequently also just an engineer [...]. Since we can allegedly make its construction principles into our own, we can only see machines wherever we look – in human cells just as in the products of nanotechnology (Nordmann, 2007b, 221).

The statement "we can only see machines wherever we look" is enlightening. Beyond the context of the quote, it tells us that the machine model, as well as the language used in engineering, is becoming a universal paradigm for describing and understanding living as well as technical objects. It applies to subcellular units and cells (Sec. 4.5), to plants and animals (Chap. 5), to genetic engineering (Chap. 6), and to humans (Sec. 7.2), but also to robots, algorithms, software agents on the Internet, autonomous cars, and artificial intelligence (Chap. 8). The machine paradigm and its related language is a kind of new *lingua franca* for understanding and shaping the world, the Latin of talking about worldviews in modern times. The model of the machine is indeed the bridge between living and technical objects, as was presumed earlier (Sec. 2.3).

Diagnosing recent developments at the crossroads of life and technology in this manner needs some warnings against possible misunderstanding. Most importantly, it has to be emphasized that describing living entities such as animals within the same machine paradigm as robots or power plants does not imply assigning them the same moral status (Sec. 9.5). Modeling, e.g., animals or humans *as* machines is different by category from saying that animals or humans *are* machines. It is extremely important to differentiate between the entities as such, the models, and the language used to describe them, and assignments of moral status and possible rights (Sec. 9.6). Modeling parts of humans, e.g., the brain, following paradigms taken from computer science, does not imply that we identify the brain as a mere computer.[4] It only works in an "as if" perspective (Gutmann and Knifka, 2015, 70). We can model living entities *as if they were* machines to meet particular purposes in an epistemological sense without assuming that they *are* machines in an ontological sense (Gutmann and Knifka, 2015).

As a conclusion, which holds for all the case studies presented in this book, we see that the techno-perspective on nature is dominant. Nature is not the model for designing new technology, but technology delivers models of how to investigate and understand living entities (option 2 above).[5] The end of nature, therefore, does not only consist of the human impact on all elements of "the grown" on planet Earth but also relies on a deep shift of understanding: from regarding nature and life as an area separated from culture and technology to a field structured according to models provided by science and technology. This shift may yet have even stronger impacts than the observable impact of human civilization on all areas of life and nature mentioned above (Sec. 9.2).

Accepting the end of nature in its traditional understanding need not collapse into nostalgia. Instead, new understandings and perspectives can be explored, which observe and understand ongoing developments at the crossroads between life and technology and between humans and nature but also transcend them with respect to human culture and perception. New understandings must not be reduced to naturalistic and biological models and perceptions of living entities. Other perspectives, and even romantic ideas, still have a place, e.g., by considering nature as a kind of garden where humans take responsible care of a more peaceful co-existence

---

[4]I concede that this differentiation is often ignored or neglected, in particular in statements by natural scientists. This frequently causes misleading understandings and unnecessary concerns in their audience (Janich, 2009; Gutmann and Knifka, 2015).

[5]This diagnosis leaves open a space to realize perhaps some of the expectations concerning the reconciliation of technology and life mentioned above. However, pursuing those effects also needs the technological view on nature in order to make learning for engineering purposes possible.

between different forms of life on Earth, with possible new, e.g., technically produced inhabitants, influenced and created by humans. Another well-known image is the utilitarian view on the new nature as a mere reservoir for human exploitation, which forms part of the Baconian understanding of human dominion over nature (Sec. 9.1); this shows itself by, e.g., modifying animals according to utilitarian human purposes (Chap. 5). In spite of criticisms that the Baconian approach is among the roots of the current ecological crisis, this utilitarian attitude is still extremely powerful. A third approach could conceptualize humans as controllers of further evolution. This view shows some similarity to the model of the gardener mentioned above but sounds much less friendly. The issue of "control" associates power and exploitation while the notion of the "gardener" focuses on care and responsibility. While the image of the controller is related to desires for full certainty, safety, and security, it ignores uncertainty, chance events, and tragic accidents. While nature and natural life are fuzzy, unplannable, and surprising, technical regulation and control aims at eliminating tragic elements by calculation (cp. the ethical dilemma situations in autonomous technology, Sec. 8.3). In accepting the imperative of control and security as dominant, the end of nature implies accepting the dominion of calculation over freedom and responsibility (cp. Sec. 8.4 at the occasion of robot rights).

Hence, the end of nature is ambivalent in itself. Abandoning chance events, e.g., in targeted genome editing, promises predictable results without surprise or disappointment, at least in the ideal case of the absence of unintended effects. However, history shows first that the absence of unintended side effects cannot be guaranteed. Second, human creativity and freedom again and again produce unpredicted and unpredictable surprises, which inherently bear the possibility of failure or disappointment. Full control, one the one hand, and freedom and creativity, on the other, build a fruitful configuration, which is, however, full of tensions. The end of nature and the transformation of purely natural evolution toward a co-evolution of nature and human-made technology shows a range of lights and shadows, which need both further enlightenment and also public debate and awareness.

9.4 Does Autonomous Technology Live?

The creation of autonomous systems able to behave independently in an unknown environment, to take action on their own based on observations of external factors, and to develop an agenda to achieve a goal has led to remarkable results in recent years (Chap. 8). Synthetic biology (Chap. 4) also addresses the creation of autonomous systems similar to viruses, which are able to "live" in at least some sense. Are humans able to create artificial life from scratch – bearing in mind that the term "artificial life" sounds like a contradiction to the ears of many?

> In fact, if synthetic biology as an activity of creation differs from genetic engineering as a manipulative approach, the Baconian *homo faber* will turn into a creator (Boldt and Müller, 2008, 387).

The contested question is not whether humans are creators or not. Evidently, they have been creators since the emergence of humankind, and even in most theological views, this is not regarded as a problem. For example, in Christian theology, humans are seen as co-creators, initially created by God and continuing creation in the world (e.g., Matern, 2016). However, this widely uncontested notion often ends up in controversy as soon as the creation of life is considered. Life is frequently regarded as something ontologically special in Western culture. Concerns of humans "playing God" are widespread and show deep uneasiness (Sec. 4.3). Does developing and producing autonomous robots, software agents, or artificial viruses result in "artificial life," and do the outcomes "live"?

At the start of this book (Sec. 2.1), we characterized "life" as a term of reflection on the criteria which objects should fulfill in order to be subsumed under the category of *living* entities. Lists of those criteria mostly include the capabilities to (without claiming completeness, cp. Mayr, 1997, 22; Gutmann, 2011b):

- reproduce
- perform metabolism
- grow
- convert and store energy
- perceive environmental conditions and react
- move (internal or external)

To clarify whether the term "life" in this sense applies to robots and artefacts produced in synthetic biology, the fulfilment of the criteria mentioned has to be assessed. But first, it is important to note that this characterization of "life" does not take care of the *process* of how these entities came into being. Subsuming something under "life," or not, depends on its observable properties, independent of whether it has, e.g., been borne by other members of the same "species," or made in a factory.

The list above demonstrates that robots indeed fulfill at least some, probably most of these criteria. They are able to move, to perceive environmental conditions and react to them, and to store energy. Self-reproduction, however, seems to be a difficult issue considering humanoid robots – but when regarding autonomous agents such as social bots on the Internet or software viruses, the criterion obviously is fulfilled. These are able to reproduce, and in particular could cause damage because of this property.[6] The issue of metabolism could also be regarded as fulfilled, at least in the case of a wide interpretation of the term, as energy consumption, e.g., of autonomous cars, can be regarded as a kind of metabolism. We also can speak of autonomous software systems digesting data and producing digital excrements. However, the latter way of talking about metabolism sounds artificial. And growth is not an issue (yet). Interpreting the production of a humanoid robot or an autonomous car as "growth" would be highly artificial compared to any common sense understanding of growth in the realm of "life." It seems that many, although not all, of the familiar attributes of "life" are at least partially fulfilled by modern autonomous technology available today. But does this technology really *live*, and what would a positive answer mean? This question is at the core of many debates:

The notion "fabrication of life" is itself an example of this ambivalence, because it raises not only fundamental doubts about whether life could be the object of fabrication but also the ontological question of whether the entities of synthetic biology [this also holds for other objects such as autonomous robots, A.G.] are living beings or artefacts (Funk *et al.*, 2019, 177).

---

[6]As a reminder: The reproducibility of nanorobots was among the major concerns with respect to nanobiotechnology and synthetic biology at the early stage of the nanotechnology debate and became the main reason for categorizing nanotechnology as the "ultimate catastrophe" (Dupuy, 2005; cp. Sec. 4.3).

As is often the case, the word "really" in the question "Does this technology *really* live?" causes immense problems. According to the understanding of "life" described above, this question does not even make sense. It is up to humans to apply certain criteria and assess whether these are fulfilled. What "really" is the case is of no interest in this *epistemological* view. This perspective, however, meets competing *ontological* understandings of "life," which, e.g., postulate an inherently necessary link between "life" and growth based on an autonomous drive. Based on this conviction, scholars conclude: "We can reject the thesis that man [*sic*] can produce "life" as an artifact by referring to the fact that aggregation is not growth and *gestalt* is not habitus" (Karafyllis, 2006, 555). Taking this dictum seriously would make it easy to clear away the problem of ethical challenges regarding how to deal with artificial life, because the latter would be a sheer impossibility. If artificial life were not possible, then there couldn't be any ethical challenges related to it.

However, the dictum that creating artificial "life" is impossible for reasons of principle must be doubted. Empirically, it seems counter-intuitive with regard to the human capacity to construct at least something similar to artificial *living beings.* These show most of the attributes of familiar criteria applying to living systems, as we saw above. Even if "a process of life cannot be reduced to chemical processes that have been linked together" (Psarros, 2006, 594), we cannot forever exclude the possibility that such linking of individual functions could result in objects that show all the properties of life according to the list above. The case studies in this book have demonstrated that there is a wide continuum of technical interventions into living systems, ranging from minimal to strong manipulations, some of which lead to largely modified or entirely new entities. Proponents of the impossibility of creating "life" would have to identify the position on this continuum which can never be transgressed by further technological progress. It is difficult and perhaps impossible to imagine that this can succeed. Therefore, a proof of the in-principle impossibility of artificial life cannot be given. The statement of the impossibility remains an implication of the ontological understanding of life, assuming an inherent drive and the power of living systems, which is regarded as inaccessible to human creation. Without sharing this ontological dogma, the argument breaks down.

As a consequence, "life" disappears as a specific ontological category (Funk *et al.*, 2019), while "living" entities remain. The constructivist epistemology is the winner against ontology (Gutmann, 2011b). Humans have constructed new entities, to which we can assign the attribute "living," or not. This renders alarmist reactions on the creative potential of biology, here at the occasion of synthetic biology (Chap. 4) misleading:

> Additionally, synthetic biology forces us to redefine "life." Is life in fact a cascade of biochemical events, regulated by the heritable code that is in (and around) the DNA and enabled by a biological machinery? Is the cell a bag of biological components that can be redesigned in a rational sense? Or is life a holistic entity that has metaphysical dimensions, rendering it more than a piece of rational machinery (de Vriend, 2006, 11)?

Again, the trap of ontologization seeking answers to the question, "What is life?," through the meaning of the noun "life" leads to absolutely unhelpful positions, which only obscure the pressing questions of assigning responsibilities and exerting them. This trap is circumvented by applying the epistemological perspective (Sec. 2.1). This insight is not that new. It goes back to many reflections in the philosophy of the sciences, in particular of biology, e.g., by German philosopher Immanuel Kant and his successors (e.g., Gutmann and Knifka, 2015). According to that legacy, the emergence of new crossroads between life and technology adds illustrative material to the fundamental change that occurs when we shift our understanding of "life" from the ontological to the epistemological level. From this perspective, the dissolution of formerly strict boundaries between life and non-living technology challenges the familiar order humans have applied to all the entities in the world. It does not, however, affect any assumed sacred "natural" and ontological order, because the existence of such an order is not presupposed in epistemological reasoning.

Finally, even hypothetically assuming that the thesis of the impossibility of artificial life were correct, it would be *irrelevant* in a practical sense. Even if the term "life," or the attribute "living," could not be attributed to artificially created objects in robotics or in synthetic biology, we would still face the question of how to deal with them responsibly in ethical respects. For example, if synthetic biology created something artificial, similar to a virus, which was able to reproduce and to spread, serious challenges to responsibility and risk management would emerge completely independent from its classification as "living" or "not living." This is similar to the case of autonomous robots: as soon as humans deliver some autonomy to technology, ethical issues of control, responsibility, and dilemmas occur, independent from classifying those robots as mere things or as something "living." *The assignment of the attribute "living" to new objects does not play any role in handling the ethical questions involved.* There is widespread concern that assigning the attribute "living" would instantaneously damage ethical standards and confuse responsibility assignment:

> [...] conceiving of technologies as living things might be just as troubling because it takes responsibility away from the people who are using those technologies and places it in the metaphoric hands of the technologies themselves, absolving us of accountability for the acts performed (Ceccarelli, 2018, 33).

However, this constitutes a false alarm and a misleading message. As soon as self-organizing and autonomous objects similar to viruses, nanorobots, or autonomous cars enter human life-worlds, new challenges with respect to responsibility are posed, independent from assigning the attribute "living" to those objects. This section in particular therefore provides us with a crucial insight for the ethics of responsibility, at least in a consequentialist respect: in searching for responsible ways of harvesting the benefits of new crossroads between life and technology as well as of dealing responsibly with risk, the assignment of the attribute "living" does not make a difference.

Analogously, this holds for the assignment of rights, welfare claims and responsibility. These assignments cannot be made to "life" in an abstract manner but rather depend on the *kind* of living entity addressed. Obviously, this will be done in an entirely different manner for viruses, plants, animals, and perhaps robots. It does not make sense to assign an inherent value or dignity to "life" in general (Link, 2010). While the carriers of value, dignity, and rights can only be living entities, their assignment depends on the respective kind of entity. For example, viruses and bacteria belong to the realm of "life" without being assigned special value or dignity (Link, 2010). In the epistemological view on "life" the assignment of the attribute "living" does not have any immediate ethical implication. These only enter the field in a second step of talking about either responsibility or moral status (see below). In a sense, this observation brings excited and exaggerated conflicts about "creating life" involving outrage and indignation down to Earth. It is also an important further step toward building the results of the reflections provided in this book into the form of the "ladder of life," presented at the end of this chapter.

9.5 Moral Status for Engineered and Created "Life"

Assigning moral status to new or heavily modified objects classified as "living" needs special consideration compared to consequentialist issues of risk and benefit. As soon as a moral status which differs from, e.g., regarding a robot as a mere thing is assigned to any technological object showing some characteristics of life, then deontological issues of rights or of inherent value could occur, depending on the kind of moral status attached. To illustrate the particular issues involved, a thought experiment introduced in Chapter 5 will be recalled. Let us assume that an artificial cat that is the functional equivalent of natural cats could be produced in a laboratory. It would exhibit all the signs of feeling comfortable and of suffering which are well known from familiar cats. Is such a technological cat entitled to the protections afforded by animal ethics and the corresponding legal regulations, or is it a technological object like a machine, which, for example, could be scrapped if the owner wants to get rid of it? The decisive question is whether the cat can "genuinely" suffer if tortured, or whether signs of pain are merely simulated by its software, which was programmed by humans. Would it be a cat or a technical simulation of a cat? Does the artificial cat live?

According to the epistemological understanding of "life" presented in Sec. 2.1 (see also above) the answer to the latter question is clear: the artificial cat does "live" because it fulfills all the criteria living entities should fulfill (Mayr, 1997; cp. Sec. 2.1). In this thought experiment, the natural and the technical cat show exactly the same behaviors. There is no alternative to assigning the attribute "living" with respect to the observable properties such as growth and self-organization (cp. the complete list in Sec. 2.1), because we cannot observe any differences between the original and the simulated cats.[7] Whether the artificial cat "really" is a cat does not matter from this phenomenological perspective. However, assigning the attribution "living" to both cats does not automatically lead to assigning them the same *moral* status. The assignment of the attribute "living" to new and created entities such as artificial viruses or autonomous agents such as robots (or artificial cats) is categorically different from assigning moral status to those entities.

Assigning moral status needs to employ criteria additional to the criteria of "living." As a result of the case studies we can identify two completely different mechanisms. Currently, most people would intuitively say that a cat is a cat because it has descended from other cats and should therefore enjoy animal protection rules and the ethos of animal welfare. Hinting at ancestry is the familiar way of assigning moral status. Human rights are rights of all humans *as humans* by their ancestry, independent of having (or not) certain properties or abilities. Human ancestry is the foundation of all regulations based on human rights. Humans who possess incompletely developed, compromised, or lost properties (e.g., premature babies, the severely disabled, or coma patients) are afforded full human rights simply because of their human ancestry.

This principle, however, comes under pressure regarding artificial life in the sense mentioned above. Even if the artificial cat, which was produced in a laboratory, shows all the properties and capabilities of a "natural" cat, it was not borne by a cat. Following the traditional principle, it thus would not enjoy animal protection rules but rather could be treated as a mere object similar to a washing machine or a computer. However, in the thought experiment, if this artificial cat showed the same behavior as normal cats, many people would experience some confusion. The question then is: do the properties shown by an artificial living entity legitimately motivate us assigning specific moral status to it, e.g., that of an animal, or perhaps a new moral status yet to be invented? In order to sharpen the alternatives the choice is between:

---

[7]This situation is similar to the "Chinese room" though experiment by Searle, which was used in Chapter 8.

(1) Applying the criterion of ancestry, which implies denying a moral status to artificially-produced living entities. It can be straightforwardly applied but might be counter-intuitive (think of the artificial cat) and could create confusion. The criterion of human ancestry for assigning rights has been criticized, e.g., by Singer ([1975] 1990), as egoistic speciesism and unfair exertion of power against other species.

(2) Applying the criterion of possessing certain abilities is more flexible. However, it implies that some humans without those abilities (e.g., severely disabled persons) would potentially not receive the moral status of humans and could not enjoy full human rights, despite being human by ancestry. Wolbring (2006) has criticized this "ableism," which increasingly makes the presence of certain capacities a precondition for someone to be recognized as fully human.

(3) Applying a situation-dependent mixture of both criteria to choose the respective "no regret" option, to be on the safer side. This seemingly inconsistent approach means applying the criterion of ancestry as established but accepting the criterion of capabilities in cases where there is no ancestry. This approach could solve the problem of the artificial cat and simultaneously avoid the problem of denying, e.g., full human rights to those who do not have all the capabilities most humans have.

Two arguments support the assignment of moral status according to the ancestry of an object rather than to its observable properties. This first operates by applying analogues. Using the familiar example, does the artificial cat "really" experience pain while tortured, and is it "really" hungry if not fed? Is a humanoid robot which shows many similarities to human behavior "really" sad if it is crying and producing tears? Is it "really" joyful if its favorite baseball team wins the match? Or are these seemingly human reactions and emotions the mere result of calculation, of algorithms produced by a software company, in order to achieve particular effects, as is always the case with technology (cp. Sec. 2.2)? Mostly, we don't have any opportunity to check what is "really" the case. Fortunately, however, we have at least some evidence. In the case of cats by ancestry, we know that they are mammals as we humans are. We can make observations and measurements providing us with good evidence that these cats can "really" suffer from pain and hunger. This is completely different with robot cats, which are not mammals but function on the basis of a source code. There is absolutely no evidence to assume that they "really" could suffer, similarly to the humanoid robot producing tears and simulating sadness.

The second argument goes deeper, to the very nature of technology. Artificial life remains technology based on a source code. Humans can look into, and understand, its mechanisms. We can also look into the genome of bacteria, plants, animals, and humans and often denote this genome as the "source code of life" (Chap. 6). However, differences remain, e.g., in the field of epigenetics. We can operate scientifically and technologically "as if" the genome were the source code similar to the source codes of apps running on computers (Gutmann and Knifka, 2015, 70). But we do not have any reason to reducing both the worlds of computers and traditionally living entities to the same model: the model of being a machine governed by some "source code." This holds in particular for humans. In this case, simulations of humans by robots remain in the realm of predictable technology as long as they do not develop self-consciousness, intentions, the capability to determine their own purposes, and their own will (Sec. 8.4). Whether this could ever be the case, is speculative. Both observations favor retaining ancestry as the dominant criterion for assigning moral status to both animals and humans.

Unfortunately, there are also arguments in the opposing direction. If the artificial and the natural cat really did show identical behaviors, then humans are in danger of becoming confused and no longer being able to distinguish properly. In a case of erroneous assignment regarding a natural cat, and subsequent denial of its animal rights and lack of compliance with animal welfare standards, irresponsible and unethical action could occur. In order to avoid such situations, it could be recommended to assign the moral status of an animal to all objects showing properties of cats, independent from their ancestry. This argument is of the "no regret" type: better to be on the safer side and assign rights to too many objects than to risk violating rights by assigning them to too few. A similar argument goes back to philosopher Immanuel Kant. His main argument for not causing pain and suffering to animals was that the way humans treat animals could influence their ways of treating humans. Violence against animals could lower the threshold for violence against humans. If humanoid robots and artificial mammals could be treated as mere things, allowing, e.g., violence, or disposability in case of dysfunction, then this treatment could influence attitudes against humans in case of their "dysfunction."

This landscape of arguments does not allow us to reach a clear conclusion based on philosophical reasoning. It demonstrates that philosophical reasoning can analyze normative uncertainties, can order, scrutinize, and classify arguments, and can draw a map of those arguments. However, it cannot make a decision, according to the model of ethics introduced in Sec. 3.1. A further process of deliberation and opinion-forming, policy-making, and regulation is ahead of us, based on the debates on genetics, the life sciences, medical issues, and enhancement since the 1970s. I see some evidence for maintaining the established criterion of ancestry as a basis for assigning moral status during the next stages of development, based on the arguments in Chapters 5 (animals), 7 (humans), and 8 (artificial forms of life). In particular, the reconstruction and critique of approaches to assign autonomy, rights, and responsibility to robots (Sec. 8.4) demonstrated this evidence.

Building on that argumentation to go one step further, we can conclude: even if some moral status were assigned to robots, e.g., with respect to the pragmatic option (3) mentioned above – this would not imply also assigning responsibility to them. Moral status could mean that robots (including artificial pets) should be treated carefully and that violence against them should be avoided. This could be a measure to recognize the feelings of many people and to avoid possible unwanted social consequences of treating those artefacts as mere things. This assignment does not have any implication for the question of assigning responsibility. The latter is, as described in previous chapters, bound to freedom, which is not compatible with the basic idea of technology (Sec. 8.4). Accordingly, we see that the assignment of moral status and subsequent rights is categorically different from assigning responsibility. While the latter becomes much more complex in socio-technical configurations involving autonomous technology because of its distribution across more actors (including that technology itself), there is nothing categorically new here. This consideration opens the way to the final section, summing up, and composing the "ladder of life."

9.6   The Ladder of "Life"

At an overarching level, the case studies of the various emerging crossroads of life and technology contribute to the following observations, diagnoses, and conclusions:

- Humans have been developing and exerting strong transformative power on the realm of life on planet Earth. By means of the ongoing scientific and technological advance, this development will continue and be accelerated in the Anthropocene, fueled by the growing global science system and the strongly increasing digital power of data gathering and data mining supported by AI algorithms.
- This transformative power extends to all areas of life and covers the entire spectrum, from minimalistic to the most complex forms of "life," i.e., of living entities. It also extends increasingly over time and affects the future of life, e.g., because of deepening human impact via genetic engineering and the hereditability of its results.
- Exerting transformative power and reshaping the world, including the realm of "life," according to human values and interests, is at the core of the increasingly globalizing Western approach rooted in the Baconian paradigm. In combination with a globalized economy and science, this paradigm is the main engine of current developments, e.g., by fueling synthetic biology, researching enhanced animals, and creating demands for genome editing and markets for human enhancement technologies.
- A similar constellation is driving rapid digitalization, including making autonomous and "living" technology such as social bots on the Internet and robots in human life-worlds possible. Deepened technical intervention into living entities and empowering technology by properties of "life" meet and increasingly converge.

- While this advance and progress is welcomed in the Baconian paradigm, the other side of the coin shows itself more urgently. Many unintended side effects, the overuse of resources, the extinction of parts of life on Earth, and lack of sustainability have become well-known challenges since the 1960s. This ambivalence is part of the mindset of many people, and many measures have been taken to improve the situation, and yet policies and the economy have not changed substantially.

- Human responsibility, therefore, is mounting, faced with many challenges in shaping the advance according to ethical values, and making use of the benefits while minimizing the side effects. Responsibility includes areas such as agenda-setting for science and innovation, the choice of methods, the early consideration of possible unintended effects, as well as risk assessment and governance. Responsibility assignments have to take into account the distribution of influence and accountability across various groups, e.g., scientists and researchers, engineers, policymakers, funders, authorities, regulators, the media, and the public, but also future humans and generations in an advocatory manner.

- The familiar consequentialist approach to responsibility assignment often reaches its limits at the occasion of the transformation of "life." It is frequently not applicable because too little (if any) valid knowledge about future consequences is available. Consequentialist approaches to responsibility have to be restricted to considering the "next steps" and specific activities, e.g., in synthetic biology, animal enhancement, or embedding autonomous robots into human practice. The more visionary and speculative questions often involve deep philosophical issues such as freedom, autonomy, the future of life, and the future of humans and are not accessible through this approach.

- Instead, hermeneutic approaches, digging deeper into the cultural, philosophical, and even religious meaning of those "big issues" have to be developed and applied, involving, amongst others, the philosophy of nature, anthropology, and the philosophy of technology.

This list can be taken as a summary of the insights gained, but we can take some steps further. *First*, the distinction made between responsibility assignments along the consequentialist paradigm and beyond puts emphasis on the epistemological dimension of the EEE model (Sec. 3.3.1). Often, uncertainties and controversies, e.g., in human genome editing or in giving some autonomy to robots, are not rooted in *ethical value conflicts* at all but rather in different assumptions about *what the future will bring*. The frequent call for ethics is often misleading, because the origin of the controversy lies in the cognitive and epistemological realm and not in moral conflict needing ethical reflection. Then, confusion and deep misunderstanding are pre-programmed because ethics cannot solve epistemic problems.

Second, responsibility debates in the consequentialist approach should be restricted to taking care of the respective "next steps," where at least some valid knowledge about consequences is available. Applying consequentialist ethics of responsibility to far-ranging and epistemologically void assumptions about the future of life, or the future of humans facing increasing technological supremacy, only causes problems. The outrage and indignation at the occasion of "playing God" debates in synthetic biology and "end of humanity" concerns in the field of human enhancement are results of this type of confusion. In those fields, responsibility reflection should address the production and dissemination of speculative, visionary, utopian, or dystopian stories of the future, instead of naively taking these futures for granted, in the consequentialist sense (Grunwald, 2017).

Third, the more speculative issues can be exploited as an important source for discussion and deliberation in order to improve mutual understanding of changing relations at the new crossroads between life and technology. Often, they address core notions of the debates, such as autonomy and morality (Chap. 6) as well as the moral status of strongly modified living entities, of hybrids, and of autonomous technology showing some properties of "life." Investigating and discussing these issues beyond any consequentialist assumptions can contribute to clarifications at the normative level, e.g., how much and what kind of autonomy do humans want to transfer to autonomous technology, and how could control be managed? Deepening understanding in these respects needs application of the hermeneutic perspective to the cultural, anthropological, and religious backgrounds and roots of related concerns, expectations, fears and hopes, values and worldviews.

Fourth, clear and reflected language is crucial for adequate understanding of the changes at the various crossroads between life and technology. Otherwise, language can be a major source of misunderstandings and can create artificial problems. Examples are the confusion between the ontological meaning of "life" and an epistemological meaning (Chap. 4); talking about robots in an anthropomorphic manner, assuming they "make decisions" and act "autonomously" (Chap. 8), and seeing humans techno-morph as machines by dropping the methodologically necessary "as if" step (Chap. 7). Emphasis must be put on the role of language, in particular with respect to the presence of technomorphic (Chaps. 4, 6, and 7) and anthropomorphic (Chap. 8) metaphors in many of the debates at the crossroads between life and technology. Metaphors often cause severe misunderstanding and confusion (Gutmann and Knifka, 2015; Funk *et al.*, 2019, 180). Postulates for a more enlightened and differentiated view on new crossroads between life and technology must therefore go hand-in-hand with care about the use of language. This leads to the final step to be taken in this book: presenting a proposal for enhancing transparency in the debates on the future of life, including humans and human responsibility.

These reflections have demonstrated that different understandings of "life," "responsibility," "autonomy," "moral status," "machine," and so forth are used in the ongoing public, scientific, and philosophical debates at the crossroads between life and technology. Often these are centered around combinations such as "engineering life," "fabricating life," or "artificial life." These understandings can be ordered along different cognitive interests of talking about "life." They form a "ladder of life." Or better: a ladder of "life" with four steps (Figs. 9.1 and 9.2). The notion of a ladder of "life" instead of a "ladder of life" points to a remarkably sensitive issue of understanding: while the "ladder of life" invokes an ontological understanding of "life" and promises a hierarchical order of living entities, the ladder of "life" addresses different levels of language according to different cognitive interests.

Culture: Assigning meaning beyond classification of something as living: talking about the "miracle of life," religious meanings of "life," worldviews, and the assumed ontological order of the world	
⇩ ⇧	⇩ ⇧
Ethics (a): Assigning moral status and rights to entities subsumed under "life," such as plants, animals, humans, robots, and hybrids	**Ethics (b):** Assigning responsibility to living entities assumed to be able to take responsibility, i.e., currently only to humans
⇩ ⇧	⇩ ⇧
Theory: Reflection on the general criteria for assigning the attributes "living" or "alive" to observed entities – "life" as a term of reflection	
⇧	⇩
Practice: Assignment of the attributes "living" or "alive" to observed entities such as bacteria, animals, robots, hybrids according to the fulfilment of criteria	

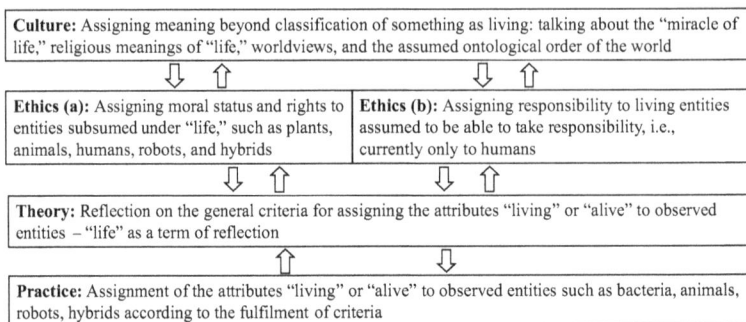

Figure 9.1   The ladder of "life" (full presentation).

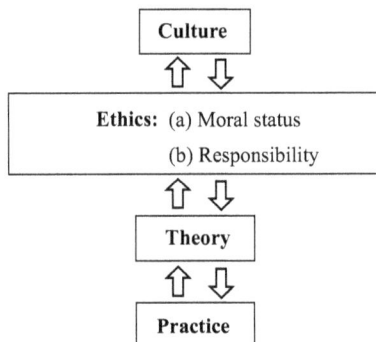

Figure 9.2   The ladder of "life" (condensed presentation).

In this ladder of "life," we distinguish ways of talking about "life" along four steps of the ladder (from the bottom to the top):

(1) *The human practice* of classifying objects and entities as "living" or "not living," whether they are natural or human-made, which is an issue in the life sciences (e.g., synthetic biology), the digital world (e.g., artificial intelligence and robots), and the human life-world. These classifications are based on criteria such as performing metabolism, growth, and reproduction.

(2) *Reflection* on and theoretical consideration of the criteria for classifying something as "living." This is the task of the theory of, e.g., biology and the philosophy of science.

(3) *Consideration* of human manipulation of entities identified as living and the consequences for the living environment, human society, and future generations with regard to emerging normative uncertainties. The ethical perspective dealing with such uncertainties comprises (a) raising the question of the moral status of and the assignment of rights to entities classified as "living," and (b) anticipating consequences of manipulations or the creation of living entities and reflecting the assignment and distribution of *responsibility*.

(4) *Hermeneutic investigation* of cultural meanings assigned to "life." "Life" is a predicate but also an expression of culture and human self-understanding. Humans talk about "life" beyond steps 1–3, e.g., by enjoying the "miracle of life," by putting "life" into a religious context, or by debating the world order and related worldviews. In debating these issues, "life" is used as a *metaphor*, and its uses can be (Funk *et al.*, 2019) investigated and analyzed through hermeneutics, philosophy of nature, anthropology, and philosophy of technology.

This ladder does not offer an ontological ordering of the living part of the world but rather maps our ways of talking about life and enables us to take care of the different levels of language involved. When talking about "life" and new crossroads with technology, we can move up and down the ladder. We can debate, classify, assign, argue, draw into doubt, and deliberate on each of the steps, following different cognitive interests and applying different perspectives on "life." At a meta-level, this ladder sums up the insights of this book gained at the occasion of the case studies, oriented to Chapters 2 and 3 on the basic notions which dominate and penetrate all the debates around "living technology" and "engineering life." By contrasting this ladder with a statement from the early debates on synthetic biology, we can clearly see the origin of confusion.

Additionally, synthetic biology forces us to redefine "life." Is life in fact a cascade of biochemical events, regulated by the heritable code that is in (and around) the DNA and enabled by a biological machinery? [...] Or is life a holistic entity that has metaphysical dimensions [...] (de Vriend, 2006, 11)?

In this statement, the author presupposes that "life" is just a thing, otherwise the question, "What is life?," would not make sense. Metaphorically speaking, this view on life corresponds to a ladder with only one step, ignoring the diversity and richness of the notion of life included in the four-step ladder presented above (Fig. 9.1). "Life" can be perceived from many perspectives, according to various levels of human practice (Fig. 9.2 shows a condensed version of the ladder of "life").

The arrows signifying travel between the steps of the ladder do not denote a logical inductive or deductive scheme of conclusion, in either direction. For example, the assignment of the attribute "living" to some entity such as a robot or an artificial virus does not imply also assigning it a moral status (Chap. 4). Each of the arrows is part of a complex structure of assignments of attributes such as "living," moral status, responsibility and cultural meaning. Therefore, the ladder is a *constructivist* picture. A series of arrows from the bottom to the top or *vice versa* is not a linear structure. A step can be bypassed. For example, if some self-organizing entities are created by using AI technology, then the issue of responsibility has to be addressed independent from whether they would be regarded as living or not. Steps closer to the bottom can be irrelevant to a step closer to the top. The ladder is not a hierarchical pyramid but offers a framework for a structured and reflected movement between the steps.

"Living technology" and "crossroads between life and technology" are the major themes of this book and parts of its title. After performing the journey through several case studies that sit at the crossroads, we arrive at a complex picture. Humankind is currently experiencing a time full of new opportunities and promising pathways to the future but is simultaneously confronted with severe challenges and uncertainties. The ladder of "life" does not provide simple answers about the future, about limits to human intervention into life, or about the future of humankind. It provides a framework for debating and shaping this future. It gives some orientation to human deliberation and decision-making. As probably the only species on planet Earth able to reflect, assign, and take over this responsibility, we are in charge of carefully and responsibly guiding the future relationship between humans, technology, and nature.

Bibliography

Abbott, A. (2006): Neuroprosthetics: In search of the sixth sense. *Nature* 442, pp. 125–127

acatech – National Academy of Science and Engineering; German National Academy of Sciences Leopoldina; Union of the German Academies of Sciences and Humanities (eds.) (2018): *Artificial Photosynthesis.* Munich: acatech

Ach, J.; Siep, L. (2006) (eds.): Nano-Bio-Ethics: Ethical and Social Dimensions of Nanobiotechnology. Berlin: Lit-Verlag

Adamatzky, A.; Komosinski, M. (2009): *Artificial Life Models in Hardware.* New York, NY: Springer

Adli, M. (2018): The CRISPR tool kit for genome editing and beyond. *Nature Communications* 9(1), p. 1911

Akcakaya, P.; Bobbin, M. L.; Guo, J. A.; *et al.* (2018): In vivo CRISPR editing with no detectable genome-wide off-target mutations. *Nature* 561(7723), pp. 416–419

Al-Balas, Q. A.; Dajani, R.; Al-Delaimy, W. K. (2019): CRISPR twins: An Islamic perspective. *Nature* 566(7745), p. 455

Albrecht, S.; Sauter, A.; König, H. (2020): *Genome Editing am Menschen.* Berlin: Office of Technology Assessment (in press)

Alpaydin, E. (2004): *Introduction to Machine Learning.* Cambridge, MA: MIT Press

Alpern, K. D. (1993): Ingenieure als moralische Helden. In: Lenk, H.; Ropohl, G. (eds.): *Technik und Ethik.* Stuttgart: Reclam, pp. 52–65

Ammann, D.; Hilbeck, A.; Lanzrein, B.; *et al.* (2007): Procedure for the implementation of the precautionary principle in biosafety commissions. *Journal of Risk Research* 10(4), pp. 487–501

Anders, G. (1964): *Die Antiquiertheit des Menschen.* München: Beck

Anderson, M.; Anderson, S. L. (eds.) (2011): *Machine Ethics*. Cambridge, MA: Cambridge University Press

Anderson, W. F. (1972): Genetic therapy. In: Hamilton, M. (ed.): *The new genetics and the future of man*. Grand Rapids, MI: Eerdmans, pp. 109–124

Araujo, M. de (2017): Editing the genome of human beings: CRISPR-Cas9 and the ethics of genetic enhancement. *Journal of Evolution and Technology* 2(1), pp. 24–42

Asdal, K.; Brenna, B.; Moser, I. (2007): *Technoscience: The Politics of Interventions*. Akademika Publishing

Asfour, T.; Kaul, L.; Wachter, M.; *et al.* (2018): ARMAR-6: A collaborative humanoid robot for industrial environments. In: *2018 IEEE-RAS 18th International Conference on Humanoid Robots (Humanoids)*. Piscataway, NJ: IEEE, pp. 447–454

Asimov, I. (1950): *I, Robot*. New York, NY: Gnome Press

Attfield, R. (1995): Genetic engineering: Can unnatural kinds be wronged? In: Wheale, P.; McNally, R.: *Animal Genetic Engineering: Of Pigs, Oncomice and Men*. London: Pluto Press, pp. 201–210

Augustine of Hippo (397): *Confessiones XI*, 20. https://en.wikisource.org/wiki/The_Confessions_of_Saint_Augustine_(Outler) [Accessed 28 Sep. 2020]

AVMA – American Veterinary Medical Association (2020): Principles of Veterinary Medical Ethics. https://www.avma.org/sites/default/files/resources/2014S_Resolution8_Attch1.pdf [Accessed 6 Feb. 2020]

Bainbridge, W. S. (2004): Progress toward cyberimmortality. In: Immortality Institute (ed.): *The Scientific Conquest of Death*. Buenos Aires: Libros en Red, pp. 107–122

Balaban, T.; Buth, G. (2005): Biomimetische Lichtsammlung. *FZK-Nachrichten* 37(4), pp. 204–209

Ball, P. (2005): Synthetic biology for nanotechnology. *Nanotechnology* 16, R1–R8

Banks, W. P.; Pockett, S. (2007): Benjamin Libet's work on the neuroscience of free will. In: Velmans, M.; Schneider, S. (eds.): *The Blackwell Companion to Consciousness*. Malden, MA: Blackwell

Bateman, S.; Gayon, J.; Allouche, S.; Goffette, J.; Marzano, M. (eds.) (2015): *Inquiring Into Animal Enhancement: Model or Countermodel of Human Enhancement?* Basingstoke, Hampshire: Palgrave Macmillan

Baumann, M. (2016): CRISPR/Cas9 genome editing – new and old ethical issues arising from a revolutionary technology. *Nanoethics* 10(2), pp. 139–159. https://doi.org/10.1007/s11569-016-0259-0

Baylis, F. (2019): Human genome editing: Our future belongs to all of us. *Issues in Science and Technology* 35(3), pp. 42–44

Beauchamp, T. L.; Childress, J. F. (2013): *Principles of Biomedical Ethics*. New York, NY: Oxford University Press

Bechtel, W.; Williamson, R. C. (1998): Vitalism. In: Craig, E. (ed.): *Routledge Encyclopedia of Philosophy*. London: Routledge, pp. 639–643

Beck, U. (1986): Risikogesellschaft. Auf dem Weg in eine andere Moderne. Frankfurt am Main: Suhrkamp. English version: Risk Society: Towards a New Modernity. London: Sage, 1992

Beck, U. (1988): *Die organisierte Unverantwortlichkeit*. Frankfurt am Main: Suhrkamp

Begley, S. (2019): After "CRISPR babies," international medical leaders aim to tighten genome editing guidelines. STAT, 24 Jan. 2019. https://www.statnews.com/2019/01/24/crispr-babies-show-need-for-more-specific-rules/ [Accessed 6 March 2019]

Benner, S. A.; Sismour, A. M. (2005): Synthetic biology. *Nature Reviews Genetics* 6(7), pp. 533–543

Berg, P.; Baltimore, D.; Brenner, S.; Roblin III, R. O.; Singer, M. F. (1981): Summary statement of the Asilomar conference on recombinant DNA molecules. *Proceedings of the National Academy of Sciences of the United States of America* 72(6), pp. 1981–1984

Bhutkar, A. (2005): Synthetic biology: Navigating the challenges ahead. *Journal of Biolaw & Business* 8(2), pp. 19–29

Blankenship, R. (2014): *Molecular Mechanisms of Photosynthesis*. 2nd edition, Oxford: Wiley

Böhle, K.; Bopp, K. (2014): What a vision: The artificial companion. A piece of vision assessment including an expert survey. *Science, Technology & Innovation Studies (STI Studies)* 10(1), pp. 155–186

Böhm, I.; Ferrari, A.; Woll, S. (2018): Visions of in vitro meat among experts and stakeholders. *Nanoethics* 12(3), pp. 211–224

Boldt, J. (2013): Life as a technical product: Philosophical and ethical aspects of synthetic biology. *Biological Theory* 8, pp. 391–401

Boldt, J. (2018): Machine metaphors and ethics in synthetic biology. *Life Sciences, Society and Policy* 14(1), pp. 1–13

Boldt, J.; Müller, O. (2008): Newtons of the leaves of grass. *Nature Biotechnology* 26(4), pp. 387–389

Boldt., J. (2014): Biotechnology, modes of action, and the value of life. In: Giese, B.; Pade, C.; Wigger, H.; von Gleich, A. (eds.): *Synthetic Biology: Character and Impact.* Heidelberg: Springer, pp. 235–248

Boldt., J. (ed.) (2016): *Synthetic Biology: Metaphors, Worldviews, Ethics, and Law.* Wiesbaden: Springer VS

Boogerd, F.; Bruggman, F.; Hofmeyer, J.; Westerhoff, H. (eds.) (2007): *Systems Biology: Philosophical Foundations.* Amsterdam: Elsevier

Börner, F.; Kehl, C.; Nierling, L. (2017): *Opportunities and Risks of Mobile and Digital Communication in the Workplace.* TAB-Fokus no. 14. https://www.tab-beim-bundestag.de/en/research/u20000.html [Accessed 7 July 2020]

Böschen, S.; Kratzer, N.; May, S. (eds.) (2006): *Nebenfolgen. Analyse zur Konstruktion und Transformation moderner Gesellschaften.* Weilerswist: Velbrück

Braun, M.; Schickl, H.; Dabrock, P. (eds.) (2018): *Between Moral Hazard and Legal Uncertainty: Ethical, Legal and Societal Challenges of Human Genome Editing.* Wiesbaden: Springer VS

Breckling, B.; Schmidt, G. (2014): Synthetic biology and genetic engineering: Parallels in risk assessment. In: Giese, B.; Pade, C.; Wigger, H.; von Gleich, A. (eds.): *Synthetic Biology. Character and Impact.* Heidelberg: Springer, pp. 197–212

Brenner, A. (2007): Leben. Eine philosophische Untersuchung. Bern: BBL

Brey, P. (2009): Values in technology and disclosive computer ethics. In: Floridi, L. (ed.): *The Cambridge Handbook of Information and Computer Ethics.* Cambridge: Cambridge University Press, pp. 41–58

Brown, N.; Rappert, B.; Webster, A. (eds.) (2000): *Contested Futures. A Sociology of Prospective Techno-Science.* Burlington, VT: Ashgate Publishing

Bruce, A.; Bruce, D. (2019): Genome editing and responsible innovation, can they be reconciled? *Journal of Agricultural and Environmental Ethics* 32, pp. 769–788. https://doi.org/10.1007/s10806-019-09789-w

Bruggeman, F. J.; Westerhoff, H. V. (2006): The nature of systems biology. *Trends in Microbiology* 15(1), pp. 45–50

Buchanan, A. (2011): *Beyond Humanity: The Ethics of Biomedical Enhancement.* Oxford, New York: Oxford University Press

Budisa, N. (2012): A brief history of the "life synthesis". In: Hacker, J.; Hecker, M. (eds.): *Was ist Leben?* Stuttgart: Wissenschaftliche Verlagsgesellschaft, pp. 99–118

Bunge, M. (1966): Technology as applied science. *Technology and Culture* 7(3), pp. 329–347

Bunge, M. (1972): Towards a philosophy of technology. In: Mitcham, C.; Mackey, R. (eds.): *Philosophy and Technology: Readings in the Philosophical Problems of Technology*. New York, NY: Free Press, pp. 62–76

Bunge, M. (2003): *Emergence and Convergence: Qualitative Novelty and the Unity of Knowledge*. Toronto: University of Toronto Press. https://doi.org/10.3138/9781442674356

Campbell, I. M.; Shaw, C. A.; Stankiewicz, P.; Lupski, J. R. (2015): Somatic mosaicism. Implications for disease and transmission genetics. *Trends in Genetics* 31(7), pp. 382–392

Campbell, S. M.; Stramondo, J. A. (2017): The complicated relationship of disability and well-being. *Kennedy Institute of Ethics Journal* 27(2), pp. 151–184

Caplan, A. (2010): Life after the synthetic cell: The end of vitalism. *Nature* 465(7297), p. 423. http://hdl.handle.net/10822/515625 [Accessed 2 Sep. 2020]

Caplan, A. (2019): Getting serious about the challenge of regulating germline gene therapy. *PLOS Biology* 17(4), e3000223

Ceccarelli, L. (2018): CRISPR as agent: A metaphor that rhetorically inhibits the prospects for responsible research. *Life Sciences, Society and Policy* 14, Art. No. 24. https://lsspjournal.biomedcentral.com/articles/10.1186/s40504-018-0088-8 [Accessed 2 Sep. 2020]

Chadwick, G. (1978): *A Systems View of Planning*. Oxford: Pergamon Press

Chakrabarti, A. M.; Henser-Brownhill, T.; Monserrat, J.; *et al.* (2019): Target-specific precision of CRISPR-mediated genome editing. *Molecular Cell* 73(4), pp. 699–713

Chakrabarty, A. M. (2003): Crossing species boundaries and making human–nonhuman hybrids: Moral and legal ramiacations. *The American Journal of Bioethics* 3(3), pp. 20–21

Chalmers, D. (2011): A computational foundation for the study of cognition. *Journal of Cognitive Science*: pp. 323–357

Chatfield, K.; Morton, D. (2018): The use of non-human primates in research. In: Schroeder, D.; *et al.* (eds.): *Ethics Dumping: Case Studies from North-South Research Collaborations*. Berlin: Springer, pp. 81–90

Cheng, J.; Fleming, G. (2009): Dynamics of light harvesting in photosynthesis. *Annual Review of Physical Chemistry* 60, pp. 242–261

Chneiweiss, H.; Hirsch, F.; Montoliu, L.; *et al.* (2017): Fostering responsible research with genome editing technologies: A European perspective. *Transgenic Research* 26(5), pp. 709–713

Chopra, P.; Kamma, A. (2006): Engineering life through synthetic biology. *In Silico Biology* 6, pp. 401–410

Church, G. (2005): Let us go forth and safely multiply. *Nature* 438, p. 423

Curchman, C. W. (1968): *The Systems Approach.* New York: Dell Publishing

Claessens, D. (1993): *Das Konkrete und das Abstrakte. Soziologische Skizzen zur Anthropologie.* Frankfurt am Main: Suhrkamp

Clausen, J. (2006): Ethische Aspekte von Gehirn-Computer-Schnittstellen in motorischen Neuroprothesen. *International Review of Information Ethics* 5, pp. 25–32

Clausen, J. (2009): Ethische Fragen aktueller Neurowissenschaften: Welche Orientierung gibt die "Natur des Menschen"? In: Hildt, E.; Engels, E.-M. (eds.): *Der implantierte Mensch.* Freiburg: Karl Alber, pp. 145–168

Cleland, C. E. (2019): *The Nature of Life: Classical and Contemporary Perspectives from Philosophy and Science.* Cambridge: Cambridge University Press

Coenen, C. (2008): *Konvergierende Technologien und Wissenschaften. Der Stand der Debatte und politischen Aktivitäten zu "Converging Technologies".* TAB-Hintergrundpapier Nr. 16. Berlin: TAB. http://www.itas.kit.edu/pub/m/2008/coen08a_zusammenfassung.htm [Accessed 2 Sep. 2020]

Coenen, C.; Schuijff, M.; Smits, M.; Klaassen, P.; Hennen, L.; Rader, M.; Wolbring, G. (2009): *Human Enhancement.* Brussels: European Parliament

COGEM & Gezondheidsraad (2017): *Ingrijpen in het DNA van de mens. Morele en maatschappelijke implicaties van kiembaanmodificatie.* Bilthoven: COGEM

COGEM (2006): *Synthetische Biologie. Een onderzoeksveld met voortschrijdende gevolgen.* Bilthoven: COGEM

Collingridge, D. (1980): *The Social Control of Technology.* London: Frances Pinter

Comstock, G. (2000): *Vexing Nature? On the Ethical Case Against Agricultural Biotechnology.* Boston, Dordrecht, London: Kluiver

Cong, L.; Ran, F. A.; Cox, D.; *et al.* (2013): Multiplex genome engineering using CRISPR/Cas systems. *Science* 339(6121), pp. 819–823

Crutzen, P. (2006): *Earth System Science in the Anthropocene.* Berlin: Springer

Cussins, J.; Lowthorp, L. (2018): Germline modification and policymaking: The relationship between mitochondrial replacement and gene editing. *The New Bioethics* 24(1), pp. 4–94

Cyranoski, D. (2018): First CRISPR babies: Six questions that remain. *Nature News*, 30.11.2018. https://www.nature.com/articles/d41586-018-07607-3 [Accessed 29 Jan. 2019]

Cyranoski, D. (2019): China to tighten rules on gene editing in humans. *Nature News*, 06.03.2019. https://www.nature.com/articles/d41586-019-00773-y [Accessed 29 Oct. 2019]

Dabrock, P. (2009): Playing God? Synthetic biology as a theological and ethical challenge. *Systems and Synthetic Biology* 3, pp. 47–54

DAK – Deutsche Angestellten Krankenkasse (2009) DAK Gesundheitsreport 2009. http://www.dak.de/content/filesopen/Gesundheitsreport_2009.pdf [Accessed 7 April 2011]

Daley, G. Q.; Lovell-Badge, R.; Steffann, J. (2019): After the storm – a responsible path for genome editing. *New England Journal of Medicine* 380(10), pp. 897–899

Damschen, G.; Schönecker, D. (eds.) (2003): *Der moralische Status menschlicher Embryonen. Pro und contra Spezies-, Kontinuums-, Identitäts- und Potentialitätsargument.* Berlin: De Gruyter

Danchin, A. (2014): The cellular chassis as the basis for new functionalities: Shortcomings and requirements. In: Giese, B.; Pade, C.; Wigger, H.; von Gleich, A. (eds.): *Synthetic Biology: Character and Impact.* Heidelberg: Springer, pp. 155–172

Danish Council of Ethics (2008): *Man or Mouse? Ethical Aspects of Chimaera Research.* Copenhagen: The Danish Council of Ethics

Darwall, S. (ed.) (2002): *Consequentialism.* Oxford: Blackwell

Dautenhahn, K. (2007): Socially intelligent robots: Dimensions of human-robot interaction. *Philosophical Transactions of the Royal Society B: Biological Sciences* 362(1480), pp. 679–704

Dawkins, R. (1989). *The Selfish Gene,* 2nd edition. Oxford: Oxford University Press

de Vriend, H. (2006): *Constructing Life: Early Social Reflections on the Emerging Field of Synthetic Biology.* The Hague: Rathenau Institute

Decker, M. (1997): *Perspektiven der Robotik. Überlegungen zur Ersetzbarkeit des Menschen.* Graue Reihe, Vol. 8, Bad Neuenahr-Ahrweiler: Europäische Akademie

Decker, M. (2012): Technology assessment of service robotics: Preliminary thoughts guided by case studies. In: Decker, M. (ed.): *Robo- and Informationethics: Some Fundamentals*. Zürich, Berlin: LIT, pp. 53–88

Decker, M.; Fleischer, T. (2010): When should there be which kind of technology assessment? A plea for a strictly problem-oriented approach from the very outset. *Poiesis & Praxis* 7, pp. 117–133. https://doi.org/10.1007/s10202-010-0074-6

Decker, M.; Dillmann, R.; Dreier, T.; Fischer, M.; Gutmann, M.; Ott, I.; Spiecker, I. (2011): Service robotics: Do you know your new companion? Framing an interdisciplinary technology assessment. *Poiesis & Praxis* 8, pp. 25–44

Decker, M.; Weinberger, N.; Krings, B.; Hirsch, J. (2017): Imagined technology futures in demand-oriented technology assessment. *Journal of Responsible Innovation* 4(2), pp. 177–196

Dennett, D. (1984): *Elbow Room: The Varieties of Free Will Worth Having*. Cambridge, MA: MIT Press

Deutscher Ethikrat (2019): *Eingriffe in die menschliche Keimbahn. Stellungnahme*. Berlin: Deutscher Ethikrat. Summary in English: Intervening in the Human Germline. Executive Summary & Recommendations. Berlijn: German Ethics Council

Dewey, J. (1920): *Reconstruction in Philosophy*. New York: Henry Holt and Company

Dewey, J. (1927): *The Public and Its Problems*. New York: Henry Holt and Company

Dewey, J. (1931): Science and society. In: Boydston, J. A. (ed.): *John Dewey: The Later Works 1925–1953*. Vol. 6, Carbondale, IL: Southern Illinois University Press, pp. 53–63

DFG – Deutsche Forschungsgemeinschaft (DFG) (2004): *Tierversuche in der Forschung*. Bonn: Lemmens Verlag & Mediengesellschaft

Drexler, K. E. (1986): *Engines of Creation – The Coming Era of Nanotechnology*. Oxford: Oxford University Press

Dunbar, C. E.; High, K. A.; Joung, J. K.; Kohn, D. B.; Ozawa, K.; Sadelain, M. (2018): Gene therapy comes of age. *Science* 359(6372), eaan4672

Dupuy, J.-P. (2005): The philosophical foundations of nanoethics. Arguments for a method. Lecture at the Nanoethics Conference, University of South Carolina, 2–5 March 2005

Dupuy, J.-P. (2007): Complexity and uncertainty: A prudential approach to nanotechnology. In: Allhoff, F.; Lin, P.; Moor, J.; Weckert, J. (eds.):

Nanoethics: The Ethical and Social Implications of Nanotechnology. Hoboken, NJ: Wiley, pp. 119–132

Dupuy, J.-P.; Grinbaum, A. (2004): Living with uncertainty: Toward the ongoing normative assessment of nanotechnology. *Techné* 8, pp. 4–25; Reprinted in: Schummer, I.; Baird, D. (eds.) (2006): *Nanotechnology Challenges: Implications for Philosophy, Ethics and Society.* Singapore *et al.*: World Scientific Publishing, pp. 287–314

Durbin, P. (ed.) (1987): *Technology and Responsibility.* Dordrecht: Reidel

EASAC – European Association of Scientific Academies (2019): *Genome Editing Scientific opportunities, public interests, and policy options in the EU.* EASAC policy report 31. https://easac.eu/fileadmin/PDF_s/reports_statements/Genome_Editing/EASAC_Report_31_on_Genome_Editing.pdf [Accessed 19 May 2020]

EC – European Comission (2013): *FET Flagships: Frequently Asked Questions.* Memo. Brussels. http://cordis.europa.eu/fp7/ict/programme/fet/flagship/doc/press28jan13-02_en.pdf [Accessed 28 Jan. 2013]

EC – European Commission (2004): *Towards a European Strategy on Nanotechnology.* Brussels: European Commission

EC – European Commission (2012): *ICT – Information and Communication Technologies: Work Programme 2013.* Luxembourg: Publications Office of the European Union

EC – European Commission (2016): *Synthetic Biology and Biodiversity.* Future Brief 15, Science for Environment Policy, Produced for the European Commission DG Environment by the Science Communication Unit, UWE, Bristol

Edwards, S. (2006): *The NanoTech Pioneers: Where Are They Taking Us?* Weinheim: Wiley-VCH

EGE – European Group on Ethics in Science and New Technologies (2005): *Ethical aspects of ICT implants in the human body.* Opinion N° 20. Brussels. https://ec.europa.eu/info/publications/ege-opinions_en [Accessed 3 Sep. 2020]

EGE – European Group on Ethics in Science and New Technologies (2016). *Statement on Gene Editing.* ec.europa.eu/info/sites/info/files/research_and_innovation/ege/gene_editing_ege_statement.pdf [Accessed 3 Sep. 2020]

EK – Ethik-Kommission (2017): *Atomatisiertes und Vernetztes Fahren.* Bericht Juni. 2017. https://www.bmvi.de/SharedDocs/DE/Publikationen/DG/bericht-der-ethik-kommission.pdf?__blob=publicationFile [Accessed 31 Aug. 2018]

Engelhardt, T. (1976): Ideology and etiology. *The Journal of Medicine and Philosophy* 1(3), pp. 256–268

Engelhardt, T. (1982): The roles of values in the discovery of illnesses, diseases, and disorders. In: Beauchamp, T.; Walters, L. (eds.): *Contemporary Issues in Bioethics*. Belmont, CA: Wadsworth, pp. 73–75

Erb, T.; Zarzycki, J. (2016): Biochemical and synthetic biology approaches to improve photosynthetic CO_2-fixation. *Current Opinion in Chemical Biology* 34, pp. 72–79

Eriksson, S.; Jonas, E.; Rydhmer, L.; Röcklingsberg, H. (2018): Invited review: Breeding and ethical perspectives on genetically modified and genome edited cattle. *Journal of Dairy Science* 101(1), pp. 1–17

Ermak, G. (2015): *Emerging Medical Technologies*. London: World Scientific

ETC Group (2007): *Extreme Genetic Engineering: An Introduction to Synthetic Biology*. http://www.etcgroup.org/sites/www.etcgroup.org/files/publication/602/01/synbioreportweb.pdf [Accessed 3 May 2015]

Fahlquist, J. N. (2017): Responsibility analysis. In: Hansson, S. O. (ed.): *The Ethics of Technology. Methods and Approaches*. London: Rowman & Littlefield, pp. 129–143

Fahlquist, J. N.; Doorn, N.; van de Poel, I. (2015): Design for the value of responsibility. In: van den Hoven, J.; Vermaas, P.; van de Poel, I. (eds.): *Handbook of Ethics, Values, and Technological Design*. Dordrecht: Springer, pp. 473–490

FAO – Food and Agriculture Organization (2006): Livestock's Long Shadow – Environmental Issues and Options. http://www.fao.org/docrep/010/a0701e/a0701e00.htm [Accessed 3 Sep. 2020]

Farah, M. J.; Illes, J.; Cook-Deegan, R.; Gardner, H.; Kandel, E.; King, P. (2004): Neurocognitive enhancement: What can we do and what should we do? *Nature Reviews Neuroscience* 5, pp. 421–425

Faunce, T.; Styring, S.; Wasielewski, M. R.; *et al.* (2013): Artificial photosynthesis as a frontier technology for energy sustainability. *Energy & Environmental Science* 6, pp. 1074–1076

Ferrari, A. (2008): *Genmaus & Co. Gentechnisch veränderte Tiere in der Biomedizin*. Stuttgart: Harald Fischer Verlag

Ferrari, A. (2015): Animal enhancement: Technovisionary paternalism and the colonisation of nature. In: Bateman, S.; Gayon, J.; Allouche, S.; Goffette, J.; Marzano, M. (eds.): *Inquiring into Animal Enhancement: Model or Countermodel of Human Enhancement?* Basingstoke, Hampshire: Palgrave Macmillan, pp. 13–33

Ferrari, A. (2017): Nonhuman animals as food in biocapitalism In: Nibert, D. (ed.): *Capitalism and Animal Oppression*. Santa Barbara, CA: Praeger Press

Ferrari, A. (2018): Animal enhancement. In: Ach, J.; Borchers, D. (eds.): *Handbuch Tierethik. Grundlagen – Kontexte – Perspektiven*. Stuttgart, Weimar: Metzler

Ferrari, A.; Coenen, C.; Grunwald, A.; Sauter, A. (2010): *Animal Enhancement. Neue technische Möglichkeiten und ethische Fragen*. Bern: Bundesamt für Bauten und Logistik BBL

Fleischer, T.; Decker, M.; Fiedeler, U. (2005): Assessing emerging technologies: Methodical challenges and the case of nanotechnologies. *Technological Forecasting & Social Change* 52, pp. 1112–1121

Foladori, G. (2008): Converging technologies and the poor: The case of nanomedicine and nanobiotechnology. In: Banse, G.; Grunwald, A.; Hronszky, I.; Nelson, G. (eds.): *Assessing Societal Implications of Converging Technological Development*. Berlin: edition sigma, p. 193–216

Frankena, W. (1939): The naturalistic fallacy. *Mind* 48, pp. 464–477

Fredens, J. (2019): Total synthesis of Escherichia coli with a recoded genome. *Nature* 569(7757), 514–518. https://doi.org/10.1038/s41586-019-1192-5

Freiermuth, J. L.; Powell-Castilla, I. J.; Gallicano, G. I. (2018): Toward a CRISPR picture. Use of CRISPR/Cas9 to model diseases in human stem cells in vitro. *Journal of Cellular Biochemistry* 119(1), pp. 62–68

Freitas, R. A. Jr. (1999): *Nanomedicine, Vol. I: Basic capabilities*. Georgetown, TX: Landes Bioscience

Freitas, R. A. Jr. (2003): *Nanomedicine, Vol. IIa: Biocompatibility*. Georgetown, TX: Landes Bioscience

French, C. E.; Horsfall, L.; Barnard, D.; *et al.* (2014): Beyond genetic engineering: Technical capabilities in the application fields of biocatalysis and biosensors. In: Giese, B.; Pade, C.; Wigger, H.; von Gleich, A. (eds.): *Synthetic Biology*. Heidelberg: Springer, pp. 113–138

Friedman, B.; Kahn, P.; Borning, A. (2006): Value sensitive design and information systems. In: Zhang, P.; Galletta, D. (eds): *Human–Computer Interaction in Management Information Systems: Foundations*. New York, London: M. E. Sharpe

Fuentes, R.; Petersson, P.; Siesser, W. B.; Caron, M. G.; Nicolelis, M. A. (2009): Spinal cord stimulation restores locomotion in animal models of Parkinson's disease. *Science* 323(5921), pp. 1578–1582

Funk, M.; Steizinger, J.; Falkner, D.; Eichinger, T. (2019): From buzz to burst: Critical remarks on the term "life" and its ethical implications in synthetic biology. *Nanoethics* 13, 173–198

Gamez, D. (2008): Progress in machine consciousness. *Consciousness and Cognition* 17(3), pp. 887–910. https://doi.org/10.1016/j.concog.2007.04.005

Gannon, F. (2003): Nano-nonsense. *EMBO Reports* 4(11), p. 1007

Gayon, J. (2010): Defining life: Synthesis and conclusions. *Origins of Life and Evolution of Biospheres* 40, pp. 231–244

Gehlen, A. (1986): *Der Mensch. Seine Natur und seine Stellung in der Welt.* Wiesbaden: Aula

Gerlinger, K.; Petermann, T.; Sauter, A. (2008): *Gene Doping. Scientific Basis – Gateways – Monitoring.* Technology Assessment Report 3, Office of Technology Assessment at the German Bundestag, Berlin: TAB. http://www.tab-beim-bundestag.de/en/pdf/publications/books/gerlinger-etal-2009-124.pdf [Accessed 3 Sep. 2020]

Gianni, R. (2016): *Responsibility and Freedom: The Ethical Realm of RRI.* London: Wiley

Gibson, D. G.; Glass, J. I.; Lartigue, C.; *et al.* (2010): Creation of a bacterial cell controlled by a chemically synthesized genome. *Science* 329(5987): pp. 52–56. https://doi.org/10.1126/science.1190719

Giese, B.; Pade, C.; Wigger, H.; von Gleich, A. (2014) (eds.): *Synthetic Biology: Character and Impact.* Heidelberg: Springer

Giese, B.; von Gleich, A. (2014): Hazards, risks, and low hazard development paths of synthetic biology. In: Giese, B.; Pade, C.; Wigger, H.; von Gleich, A. (eds.): *Synthetic Biology. Character and Impact.* Heidelberg: Springer, pp. 173–196

Glees, P. (2005): *The Human Brain.* Cambridge: Cambridge University Press

Goodall, N. J. (2014): Machine ethics and automated vehicles. In: Meyer, G.; Beiker, S. (eds.): *Road Vehicle Automation: Lecture Notes in Mobility.* Cham: Springer. https://doi.org/10.1007/978-3-319-05990-7_9

Goodsell, D. S. (2004): *Bionanotechnology: Lessons from Nature.* New York: Wiley

Greely, H.; Sahakian, B.; Harris, J.; Kessler, R.; Gazzaniga, M.; Campbell, P.; Farah, M. (2008): Towards responsible use of cognitive-enhancing drugs by the healthy. *Nature* 456, pp. 702–706

Greely, H. T. (2019): CRISPR'd babies: Human germline genome editing in the "He Jiankui affair". *Journal of Law and the Biosciences* 6(1), pp. 111–183

Green, R. M. (2007): *Babies By Design: The Ethics of Genetic Choice.* New Haven, CT: Yale University Press, pp. 96–97

Grinbaum, A.; Groves, C. (2013): What is "responsible" about responsible innovation? Understanding the ethical issues. In: Owen, R.; Bessant, J. R.; Heintz, M. (eds.): *Responsible Innovation: Managing the Responsible Emergence of Science and Innovation in Society.* London: Wiley, pp. 119–142

Grolle, J. (2010): Konkurrenz für Gott. *SPIEGEL* 1/2010, pp. 110–119

Grunwald, A. (2000): Against over-estimating the role of ethics in technology. *Science and Engineering Ethics* 6, pp. 181–196

Grunwald, A. (2003): Methodical reconstruction of ethical advises. In: Bechmann, G.; Hronszky, I. (eds.): *Expertise and Its Interfaces.* Berlin: Edition Sigma, pp. 103–124

Grunwald, A. (2006): Scientific independence as a constitutive part of parliamentary technology assessment. *Science and Public Policy* 33(2), pp. 103–113. https://doi.org/10.3152/147154306781779073

Grunwald, A. (2007): Converging technologies: Visions, increased contingencies of the conditio humana, and search for orientation. *Futures* 39(4), pp. 380–392

Grunwald, A. (2008a): Nanoparticles: Risk management and the precautionary principle. In: Jotterand, F. (ed.): *Emerging Conceptual, Ethical and Policy Issues in Bionanotechnology.* Berlin: Springer, pp. 85–102

Grunwald, A. (2008b): Ethical guidance for dealing with unclear risk. In: Wiedemann, P.; Schütz, H. (eds.): *The Role of Evidence in Risk Characterization. Making Sense of Conflicting Data.* Weinheim: Wiley-VCH, pp. 185–202

Grunwald, A. (2010): From speculative nanoethics to explorative philosophy of nanotechnology. *NanoEthics* 4(2), pp. 91–101

Grunwald, A. (2012): *Responsible Nanobiotechnology: Philosophy and Ethics.* Singapore: Jenny Stanford Publishing

Grunwald, A. (2013): Modes of orientation provided by futures studies: Making sense of diversity and divergence. *European Journal of Futures Research* 2(30), pp. 1–9. https://doi.org/10.1007/s40309-013-0030-5

Grunwald, A. (2014a): The hermeneutic side of responsible research and innovation. *Journal of Responsible Innovation* 1(3), pp. 274–291. https://doi.org/10.1080/23299460.2014.968437

Grunwald, A. (2014b): Synthetic biology as technoscience and the EEE concept of responsibility. In: Giese, B.; Pade, C.; Wigger, H.; von Gleich, A. (eds.) *Synthetic Biology. Character and Impact.* Heidelberg: Springer, pp. 249–266

Grunwald, A. (2016a): Synthetic biology: Seeking for orientation in the absence of valid prospective knowledge and of common values. In: Hansson, S. O.; Hirsch Hadorn, G. (eds.): *The Argumentative Turn in Policy Analysis: Reasoning about Uncertainty.* Cham: Springer, pp. 325–344. https://doi.org/10.1007/978-3-319-30549-3_14

Grunwald, A. (2016b): *The Hermeneutic Side of Responsible Research and Innovation.* London: Wiley-ISTE

Grunwald, A. (2016c): What does the debate on (post)human futures tell us? Methodology of hermeneutical analysis and vision assessment. In: Hurlbut, J. B.; Tirosh-Samuelson, H. (eds.): *Perfecting Human Futures: Transhuman Visions and Technological Imaginations.* Wiesbaden: Springer, pp. 35–50

Grunwald, A. (2016d): Societal risk constellations for autonomous driving: Analysis, historical context and assessment. In: Maurer, M.; Gerdes, J. C.; Lenz, B.; Winner, H. (eds.): *Autonomous Driving: Technical, Legal and Social Aspects.* Berlin: Springer, pp. 641–663. https://doi.org/10.1007/978-3-662-48847-8_30

Grunwald, A. (2017): Assigning meaning to NEST by technology futures: Extended responsibility of technology assessment in RRI. *Journal of Responsible Innovation* 4(2), pp. 100–117. https://doi.org/10.1080/23299460.2017.1360719

Grunwald, A. (2018): Self-driving cars: Risk constellation and acceptance issues. *DELPHI – Interdisciplinary Review of Emerging Technologies* 1(1), pp. 8–13

Grunwald, A. (2019a): *Technology Assessment in Practice and Theory.* London: Routledge

Grunwald, A. (2019b): *Der unterlegene Mensch. Zur Zukunft der Menschheit im Angesicht von Algorithmen, Robotern und Künstlicher Intelligenz.* München: riva

Grunwald, A.; Julliard, Y. (2005): Technik als Reflexionsbegriff. Überlegungen zur semantischen Struktur des Redens über Technik. *Philosophia Naturalis* 42(1), pp. 127–157

Grunwald, A.; Julliard, Y. (2007): Nanotechnology: Steps towards understanding human beings as technology? *NanoEthics* 1, pp. 77–87

Grunwald, A.; Gutmann, M.; Neumann-Held, E. (eds.) (2002): *On Human Nature: Anthropological, Biological, and Philosophical Foundations.* Berlin: Springer

Gunkel, D. (2018). *Robot Rights.* Cambridge, MA: MIT Press

Gutmann, M. (2004): *Erfahren von Erfahrungen. Dialektische Studien zur Grundlegung einer philosophischen Anthropologie.* Bielefeld: transcript

Gutmann, A. (2011a): The ethics of synthetic biology: Guiding principles for emerging technologies. *Hastings Center Report* 41(4), pp. 17–22

Gutmann, M. (2011b): Life and human life. In: Korsch, D.; Griffioen, A. (eds.): *Interpreting Religion.* Tübingen: Mohr, pp. 163–185

Gutmann, M. (2017): *Leben und Form. Zur technischen Form des Wissens vom Lebendigen.* Wiesbaden: Springer VS

Gutmann, M.; Decker, M.; Knifka, J. (eds.) (2015): *Evolutionary Robotics, Organic Computing and Adaptive Ambience.* Vienna: LIT

Gutmann, M.; Knifka, J. (2015): Biomorphic and technomorphic metaphors: Some arguments why robots do not evolve, why computing is not organic, and why adaptive technologies are not intelligent. In: Gutmann, M.; Decker, M.; Knifka, J. (eds.): *Evolutionary Robotics, Organic Computing and Adaptive Ambience.* Vienna: LIT, pp. 53–80

Gyngell, C.; Douglas, T.; Savulescu, J. (2017): The ethics of germline gene editing. *Journal of Applied Philosophy* 34, pp. 498–513

Gyngell, C.; Bowman-Smart, H.; Savulescu, J. (2019): Moral reasons to edit the human genome: Picking up from the Nuffield report. *Journal of Medical Ethics* 45(8), pp. 514–523

Habermas, J. (1970): Toward a Rational Society. London: Heinemann (Original: Technik und Wissenschaft als Ideologie. Frankfurt am Main: Suhrkamp, 1968)

Habermas, J. (1973): Wahrheitstheorien. In: Fahrenbach, H. (ed.): *Wirklichkeit und Reflexion. Walther Schulz zum sechzigsten Geburtstag.* Pfullingen: Neske, pp. 211–265

Habermas, J. (1988): *The Theory of Communicative Action.* Boston, MA: Beacon Press. German version: *Theorie des kommunikativen Handelns.* Frankfurt am Main: Suhrkamp, 1984

Habermas, J. (1992): *Between Facts and Norms: Contributions to a Discourse Theory of Law and Democracy.* Boston, MA: Beacon Press

Habermas, J. (2005): *The Future of the Human Nature.* Cambridge, MA: Polity Press. German version: *Die Zukunft der menschlichen Natur.* Frankfurt am Main: Suhrkamp, 2001

Hagen, K.; Engelhard, M.; Toepfer, G. (eds.) (2016): *Ambivalences of Creating Life. Societal and Philosophical Dimensions of Synthetic Biology.* Heidelberg *et al.*: Springer

Haggard, P. (2019): The neurocognitive bases of human volition. *Annual Review of Psychology* 70(1), pp. 9–28

Hamilton, M. (ed.) (1972): *The New Genetics and the Future of Man.* Grand Rapids, MI: Eerdmans

Hansson, S. O. (2006): Great uncertainty about small things. In: Schummer, J.; Baird, D. (eds.): *Nanotechnology Challenges: Implications for Philosophy, Ethics, and Society.* Singapore *et al.*: World Scientific Publishing, pp. 315–325

Hansson, S. O. (2010): Risk and safety in technology. In: Meijers, A. (ed.): *Philosophy of Technology and Engineering Sciences.* Vol. 9. Amsterdam: Elsevier, pp. 1069–1102

Hansson, S. O. (ed.) (2017): *The Ethics of Technology. Methods and Approaches.* London: Rowman & Littlefield

Hardman, A.; Jones, C. (eds.) (2010): *Philosophy of Sport.* Cambridge: Cambridge Scholars Publishing

Harremoes, P.; Gee, D.; MacGarvin, M.; Stirling, A.; Keys, J.; Wynne, B.; Guedes Vaz, S. (eds.) (2002): *The Precautionary Principle in the 20th Century. Late Lessons from Early Warnings.* London: Earthscan

Hartmann, D. (1996): Kulturalistische Handlungstheorie. In: Hartmann, D.; Janich, P. (ed.): *Methodischer Kulturalismus. Zwischen Naturalismus und Postmoderne.* Frankfurt am Main: Suhrkamp, pp. 70–114

Hatada, I. (ed.) (2017): *Genome Editing in Animals.* Berlin: Springer

Hatsopoulos, N. G.; Donoghue, J. P. (2009): The science of neural interface systems. *Annual Review of Neurosciences* 32, pp. 249–266

Hennen, L.; Grünwald, R.; Revermann, C.; Sauter, A. (2007): *Hirnforschung.* Berlin: TAB. https://www.tab-beim-bundestag.de/de/untersuchungen/u117.html [Accessed 4 Sep. 2020]

Hennen, L.; Nierling, L.; Hebakova, L. (2012): Parliamentary TA in Germany. In: Ganzevles, J.; van Est, R. (eds.): *TA Practices in Europe.* European Commission: Brussels, pp. 100–120

Hermann, M.; Pentek, T.; Otto, B. (2016). Design principles for Industrie 4.0 scenarios. In: *IEEE, 2016 49th Hawaii International Conference on System Sciences (HICSS).* Koloa, HI: IEEE, pp. 3928–3937. https://doi.org/10.1109/HICSS.2016.488

Hevelke, A.; Nida-Rümelin, J. (2015): Responsibility for crashes of autonomous vehicles: An ethical analysis. *Science and Engineering*

Ethics 21(3), pp. 619–630. https://doi.org/10.1007/s11948-014-9565-5

Hildt, E.; Engels, E.-M. (eds.) (2009): *Der implantierte Mensch. Therapie und Enhancement im Gehirn.* Freiburg: Alber

Höffe, O.; Honnefelder, L.; Isensee, J.; Kirchhof, P. (eds.) (2002): *Gentechnik und Menschenwürde.* Köln: DuMont

Holtug, N. (1996): Is welfare all that matters in our moral obligations to animals? *Acta Agriculturae Scandinavica Sect. A,* Animal Science Supplement 27, pp. 16–21

Hook, C. (2004): The Techno Sapiens are Coming. *Christianity Today* 48(1), pp. 36–40

Hornberg-Schwetzel, S. (2008): Therapie und Enhancement. Der Versuch einer wertvollen Unterscheidung. In: Honnefelder, L.; Sturma, D. (eds.): *Jahrbuch für Wissenschaft und Ethik.* Vol. 13. Berlin, New York: de Gruyter, pp. 207–221

Hubig, C. (2007): *Die Kunst des Möglichen II. Ethik der Technik als provisorische Moral.* Bielefeld: transcript

Hunter, D. (2013): How to object to radically new technologies on the basis of justice: The case of synthetic biology. *Bioethics* 27(8), pp. 426–434

Hurlbut, J. B.; Tirosh-Samuelson, H. (eds.) (2016): *Perfecting Human Futures: Transhuman Visions and Technological Imaginations.* Springer: Wiesbaden

Hyun, I.; Scharf-Deering, J. C.; Lunshof, J. E. (2020): Ethical issues related to brain organoid research. *Brain Research* 1732, 146653. https://doi.org/10.1016/j.brainres.2020.146653

Ilulissat Statement (2008): Synthesizing the future: A vision for the convergence of synthetic biology and nanotechnology. *Kavli Futures Symposium "The Merging of Bio and Nano: Toward Cyborg Cells,"* 11–15 June 2007. Ilulissat, Greenland

IRGC – International Risk Governance Council (2009): *Risk Governance of Synthetic Biology.* Geneva: IRGC

Ishii, T. (2017): Germ line genome editing in clinics: The approaches, objectives and global society. *Briefings in Functional Genomics* 16(1), pp. 46–56

JafariNaimi, N. (2017): Our bodies in the trolley's path, or why self-driving cars must *not* be programmed to kill. *Science, Technology & Human Values* 43(2), pp. 302–323. https://doi.org/10.1177/0162243917718942

Janich, P. (2001): *Logisch-pragmatische Propädeutik.* Weilerswist: Velbrück

Janich, P. (2002): Modelle und Modelliertes. In: Gethmann, C. F.; Lingner, S. (eds.): *Integrative Modellierung zum Globalen Wandel*. Berlin: Springer, pp. 25–48

Janich, P. (2006): Nanotechnology and the philosophy of science. In: Schmid, G.; Ernst, H.; Grünwald, W.; Grunwald, A.; *et al*. (2006): *Nanotechnology – Perspectives and Assessment*. Berlin *et al.*: Springer, pp. 13–43

Janich, P. (2009): *Kein neues Menschenbild. Zur Sprache der Hirnforschung*. Frankfurt: Suhrkamp

Jasanoff, S. (2016): Perfecting the human: Posthuman imagineries and technologies of reason. In: Hurlbut, J. B.; Tirosh-Samuelson, H. (eds.): *Perfecting Human Futures: Transhuman Visions and Technological Imaginations*. Wiesbaden: Springer, pp. 73–96

Jasanoff, S.; Hurlbut, B.; Saha, K. (2015): CRISPR democracy: Gene editing and the need for inclusive deliberation. *Issues in Science and Technology* 32, pp. 25–32

Jasanoff, S.; Hurlbut, J. B. (2018): A global observatory for gene editing. *Nature* 555(7697), pp. 435–437

Johnson, B. D. (2015): *21st Century Robot*. Sebastopol, CA: Maker Media Inc.

Jömann, N.; Ach, J. S. (2006): Ethical implications of nanobiotechnology: State-of-the-art survey of ethical issues related to nanobiotechnology. In: Ach, J.; Siep, L. (eds.): *Nano-Bio-Ethics. Ethical and Social Dimensions of Nanobiotechnology*. Berlin: Lit, pp. 13–62

Jonas, H. (1979): *Das Prinzip Verantwortung. Versuch einer Ethik für die technologische Zivilisation*. Frankfurt am Main: Insel. English version: *The Imperative of Responsibility: In Search of an Ethics for the Technological Age*. Chicago, IL: University of Chicago Press, 1984

Jones, R. A. L. (2004): *Soft Machines. Nanotechnology and Life*. Oxford: Oxford University Press

Jorqui-Azofra, M. (2020): Regulation of clinical xenotransplantation: A reappraisal of the legal, ethical, and social aspects involved. *Methods in Molecular Biology* 2110, pp. 315–358. https://doi.org/10.1007/978-1-0716-0255-3_20

Jotterand, F. (ed.) (2008a): *Emerging Conceptual, Ethical and Policy Issues in Bionanotechnology*. Berlin: Springer

Jotterand, F. (2008b): Beyond therapy and enhancement: The alteration of human nature. *Nanoethics* 2(1), pp. 15–23

Joy, B. (2000): Why the future does not need us. *Wired Magazine* 8(4), pp. 238–263

Juengst, E. T. (1997): Can enhancement be distinguished from prevention in genetic medicine? *The Journal of Medicine and Philosophy* 22, pp. 125–142

Kalds, P.; Zhou, S.; Cai, B.; Liu, J.; Wang, Y.; Petersen, B.; Sonstegard, T.; Wang, X.; Chen, Y. (2019): Sheep and goat genome engineering: From random transgenesis to the CRISPR era. *Frontiers in Genetics* 10, pp. 750–762. https://doi.org/10.3389/fgene.2019.00750

Karafyllis, N. C. (2006): Biofakte: Grundlagen, Probleme und Perspektiven. *Erwägen Wissen Ethik (EWE)* 17(4), pp. 547–558

Karafyllis, N. C. (2008): Ethical and epistemological problems of hybridizing living beings: Biofacts and body shopping. In: Li, W.; Poser, H. (eds.): *Ethical Considerations on Today's Science and Technology. A German-Chinese Approach.* Münster: LIT, pp. 185–198

Karpowicz, P.; Cohen, C. B.; van der Kooy, D. (2005): Developing human-nonhuman chimeras in human stem cell research: Ethical issues and boundaries. *Kennedy Institute of Ethics Journal* 15(2), pp. 107–134

Kehl, C. (2018): *Robotik und assistive Neurotechnologien in der Pflege – gesellschaftliche Herausforderungen.* TAB-Arbeitsbericht 177. Berlin: TAB. https://www.tab-beim-bundestag.de/de/untersuchungen/u106002.html [Accessed 14 Sep. 2020]

Khushf, G. (2007): The ethics of NBIC convergence. *Journal of Medicine and Philosophy* 32(3), pp. 185–196

King, N.; Hyde, M. J. (eds.) (2012): *Bioethics, Public Moral Argument, and Social Responsibility.* London: Routledge

Kinnebrock, A. (1997): *Künstliches Leben: Anspruch und Wirklichkeit.* Munich: Oldenbourg

Knoepfler, P. (2015): *GMO Sapiens: The Life-Changing Science of Designer Babies.* Singapore, Hackensack, NJ: World Scientific. https://doi.org/10.1142/9542

Kobayawa, K.; Kobayakawa, R.; Matsumoto, H.; *et al.* (2007): Innate versus learned odour processing in the mouse olfactory bulb. *Nature* 450(7169), pp. 503–508

Koch, C. (2004): *The Quest for Counsciousness: A Neurobiological Approach.* Englewood, CO: Roberts and Company Publishers

Kolb, B.; Whishaw, I. Q. (2013): *An Introduction to Brain and Behavior.* New York, NY: Freeman-Worth

König, H.; Frank, D.; Heil, R.; Coenen, C. (2016): Synthetic biology's multiple dimensions of benefits and risks: Implications for governance and policies. In: Boldt, J. (ed.): *Synthetic Biology. Metaphors, Worldviews, Ethics, and Law*. Wiesbaden: Springer, pp. 217–232

Kowarsch, M. (2016): *A Pragmatist Orientation for the Social Sciences in Climate Policy. How to Make Integrated Economic Assessments Serve Society*. Boston Studies in the Philosophy and History of Science, Vol. 323. Cham: Springer International

Kraft, J. C.; Osterhaus, G. l.; Ortiz, A. N.; Garris, P. A.; Johnson, M. A. (2009): In vivo dopamine release and uptake impairments in rats treated with 3-nitropropionic acid. *Neuroscience* 161(3), pp. 940–949

Kralj, M.; Pavelic, K. (2003): Medicine on a small scale: How molecular medicine can benefit from self-assembled and nanostructured materials. *EMBO reports* 4(11), pp. 1008–1012

Krimsky, S. (2019): Ten ways in which He Jiankui violated ethics. *Nature Biotechnology* 37, pp. 19–20. https://doi.org/10.1038/nbt.4337

Lander, E.; Baylis, F.; Zhang, F.; *et al.* (2019). Adopt a moratorium on heritable genome editing. *Nature* 567, pp. 165–168

Lang, A.; Spök, A.; Gruber, M.; *et al.* (2019): *Genome Editing – Interdisziplinäre Technikfolgenabschätzung*. TA-SWISS Vol. 70/2019. Zurich: vdf

Latour, B. (1987): *Science in Action: How to Follow Scientists and Engineers Through Society*. Cambridge, MA: Hardward University Press

LBNL – Lawrence Berkeley National Laboratory (2006): Homepage. http://www.lbl.gov/pbd/synthbio/default.htm [Accessed 13 Apr. 2015]

Lenk, H. (1992): *Zwischen Wissenschaft und Ethik*. Frankfurt am Main: Suhrkamp

Li, J.; Walker, S.; Nie, J.-B.; Zhang, X.Q. (2019) Experiments that led to the first gene-edited babies: The ethical failings and the urgent need for better governance. *Journal of Zhejiang University SCIENCE B* 20(1), pp. 32–38

Liang, P.; Ding, C.; Sun, H.; *et al.* (2017): Correction of β-thalassemia mutant by base editor in human embryos. *Protein & Cell* 8(11), pp. 811–822

Lin, P.; Abney, K.; Bekey, G. A. (eds.) (2012): *Robot Ethics: The Ethical and Social Implications of Robotics*. Cambridge, MA: MIT Press

Link, H.-J. (2010): In-depth analysis of outstanding philosophical issues. In: Synth-Ethics Consortium (ed.): *Synth-Ethics: Identification of Ethical Issues and Analysis of Public Discourse*. FP7 Report WP1, pp. 40–49

Liu, C. (2019): Multiple social credit systems in China. *Economic Sociology: The European Electronic Newsletter* 21(1), pp. 22–32

Lovell-Badge, R. (2019): CRISPR babies: A view from the centre of the storm. *Development* 146(3), dev175778. https://doi.org/10.1242/dev.175778

Lu, T., Yang, B., Wang, R., Qin, C. (2019). Xenotransplantation: Current Status in Preclinical Research. *Frontiers in Immunology* 10, 3060. https://doi.org/10.3389/fimmu.2019.03060

Ma, D.; Liu, F. (2015): Genome editing and its applications in model organisms. *Genomics, Proteomics & Bioinformatics* 13(6), pp. 336–344. https://doi.org/10.1016/j.gpb.2015.12.001

Ma, H.; Marti-Gutierrez, N.; Park, S.; *et al.* (2017): Correction of a pathogenic gene mutation in human embryos. *Nature* 548, pp. 413–419

Machery, E. (2012): Why I stopped worrying about the definition of life … and why you should as well. *Synthese* 185(1), pp. 145–164

Mahner, M.; Bunge, M. (1997): *Foundations of Biophilosophy*. Heidelberg: Springer

Mainzer, K. (2015): Life as machine? From life science to cyberphysical systems. In: Gutmann, M.; Decker, M.; Knifka, J. (eds.): *Evolutionary Robotics, Organic Computing and Adaptive Ambience*. Vienna: LIT, pp. 13–28

Mainzer, K. (2016): *Künstliche Intelligenz – Wann übernehmen die Maschinen?* Berlin, Heidelberg: Springer

Manzano, M. (1999): *Model Theory*. Oxford: Oxford University Press

Manzlei, C.; Schleupner, L.; Heinz, R. (eds) (2016): *Industrie 4.0 im internationalen Kontext*. Berlin: VDE Verlag

Margolis, J. (1981): The concept of disease. In: Caplan, A. (ed.): *Concepts of Health and Disease: Interdisciplinary Perspectives*. Reading, MA: Addison-Wesley, pp. 561–577

Matern, H. (2016): Creativity and technology: Humans as co-creators. In: Boldt., J. (ed.): *Synthetic Biology: Metaphors, Worldviews, Ethics, and Law*. Wiesbaden: Springer, pp. 71–86

Matern, H.; Ried, J.; Braun, M.; Dabrock, P. (2016): Living machines: On the genesis and systematic implications of a leading metaphor of synthetic biology. In: Boldt, J. (ed.): *Synthetic Biology. Metaphors, Worldviews, Ethics, and Law*. Wiesbaden: Springer, pp. 47–60

Maurer, M.; Gerdes, J.; Lenz, B.; Winner, H. (eds.) (2016): *Autonomous Driving. Technical, Legal and Social Aspects*. Heidelberg: Springer Open

Maurer, S.; Lucas, K.; Terrel, S. (2006): *From Understanding to Action: Community Based Options for Improving Safety and Security in Synthetic Biology*. Berkeley, CA: University of California

Mayr, E. (1997): *This Is Biology*. London: Belknap Press

McKenna, M.; Perelboom, D. (2016): *Free Will: A Contemporary Introduction*. New York, NY: Routledge

Merkel, R.; Boer, G.; Fegert, J.; Galert, T.; Hartmann, D.; Nuttin, B.; Rosahl, S. (2007): *Intervening in the Brain. Changing Psyche and Society*. Berlin: Springer

Miller, S.; Selgelid, M. (2006): *Ethics and the Dual-Use Dilemma in the Life Sciences*. Berlin *et al.*: Springer

Mitcham, C. (1994): *Thinking Through Technology: The Path Between Engineering and Philosophy*. Chicago, IL: University of Chicago Press

Miura, H.; Quadros, R. M.; Gurumurthy, C. B.; Ohtsuka, M. (2017): Easi-CRISPR for creating knock-in and conditional knockout mouse models using long ssDNA donors. *Nature Protocols* 13, pp. 195–215

Mnyusiwalla, A.; Daar, A. S.; Singer, P. A. (2003): Mind the gap: Science and ethics in nanotechnology. *Nanotechnology* 14, pp. R9–R13

Moniz, A. (2015): Robots and humans as co-worker? The human-centred perspective of work with autonomous systems. In: Gutmann, M.; Decker, M.; Knifka, J. (eds.): *Evolutionary Robotics, Organic Computing and Adaptive Ambience*. Vienna: LIT, pp. 147–176

Moor, J.; Weckert, J. (2004): Nanoethics: Assessing the nanoscale from an ethical point of view. In: Baird, D.; Nordmann, A.; Schummer, J. (eds.): *Discovering the Nanoscale*. Amsterdam: IOS Press, pp. 301–310

Moore, D. S. (2015): *The Developing Genome: An Introduction to Behavioral Epigenetics*. Oxford: Oxford University Press

Morar, N. (2015): An empirically informed critique of Habermas' argument from human nature. *Science and Engineering Ethics* 21(1), pp. 95–113

Moreno, A. (2007): A systemic approach to the origin of biological organization. In: Boogerd, F.; Bruggman, F.; Hofmeyer, J.; Westerhoff, H. (eds.): *Systems Biology. Philosophical Foundations*. Amsterdam: Elsevier, pp. 243–268

Moritz, C. T.; Perlmutter, S. I.; Fetz, E. E. (2008): Direct control of paralysed muscles by cortical neurons. *Nature* 456(7222), pp. 639–642

Müller, S. (2006): Minimal-invasive und nanoskalige Therapien von Gehirnerkrankungen: Eine medizinethische Diskussion. In: Nordmann, A.; Schummer, J.; Schwarz, A. (eds.): *Nanotechnologien im Kontext*. Berlin: Akademische Verlagsgesellschaft, pp. 346–370

Mulvihill, J. J.; Capps, B.; Joly, Y.; Lysaght, T.; Zwart, H. A. E.; Chadwick, R. (2017): Ethical issues of CRISPR technology and gene editing through the lens of solidarity. *British Medical Bulletin* 122(1), pp. 17–29

Munthe, C. (2018): *Precaution and Ethics: Handling Risks, Uncertainties and Knowledge Gaps in the Regulation of New Biotechnologies*. Contributions to Ethics and Biotechnology, Vol. 12. Bern: ECNH

National Academies of Sciences, Engineering, and Medicine (2017): *Human Genome Editing. Science, Ethics, and Governance.* Washington, DC: National Academies Press

Nerurkar, M. (2012): *Amphibolie der Reflexionsbegriffe und transzendentale Reflexion. Das Amphibolie-Kapitel in Kants Kritik der reinen Vernunft.* Würzburg: Königshausen & Neumann

Neumann, S. (2008): Cosmetic surgery: Customer service or professional misconduct. *Canadian Veterinary Journal* 49(5), pp. 501–504

Nick, P.; Fischer, R.; Gradl, D.; Gutmann, M.; Kämper, J.; Lamparter, T.; Riemann, M. (2019): *Modellorganismen.* Berlin: Springer Spektrum

NNI – National Nanotechnology Initiative (1999): *National Nanotechnology Initiative.* Washington, DC: Office of Science and Technology Policy

Nordenfelt, L. (1993): *Quality of Life, Health, and Happiness.* Aldershot: Avebury

Nordmann, A. (2007a): If and then: A critique of speculative nanoethics. *Nanoethics* 1, pp. 31–46

Nordmann, A. (2007b): Entflechtung – Ansätze zum ethisch-gesellschaftlichen Umgang mit der Nanotechnologie. In: Gazsó, A.; Greßler, S.; Schiemer, F. (eds.): *Nano – Chancen und Risiken aktueller Technologien.* Vienna: Springer, pp. 215–229

Nordmann, A. (2010): A forensic of wishing. Technology assessment in the age of technoscience. *Poiesis & Praxis* 7, pp. 5–15

Nordmann, A. (2014a): Synthetic biology and the limits of science. In: Giese, B.; Pade, C.; Wigger, H.; von Gleich, A. (eds.): *Synthetic Biology. Character and Impact.* Heidelberg: Springer, pp. 31–58

Nordmann, A. (2014b): Responsible innovation, the art and craft of future anticipation. *Journal of Responsible Innovation* 1, pp. 87–98

Nuffield Council on Bioethics (2005): *The Ethics of Research Involving Animals.* London: Nuffield Council on Bioethics

Nuffield Council on Bioethics (2016): *Genome Editing: An Ethical Review.* London: Nuffield Council on Bioethics. http://nuffieldbioethics. org/wp-content/uploads/Genome-editing-an-ethical-review.pdf [Accessed 14 Sep. 2020]

Nuffield Council on Bioethics (2018): *Genome Editing and Human Reproduction. Social and Ethical Issues.* London: Nuffield Council on Bioethics. http://nuffieldbioethics.org/wp-content/uploads/ Genome-editing-and-human-reproduction-FINAL-website.pdf [Accessed 14 Sep. 2020]

O'Neill, C. (2016): W*eapons of Math Destruction: How Big Data Increases Inequality and Threatens Democracy*. New York, NY: Crown

Olson, S. (ed.) (2016): *International Summit on Human Gene Editing: A Global Discussion*. Washington, DC: National Academies Press

Orwat, C.; Raabe, O.; Buchmann, E.; *et al.* (2010): Software als Institution und ihre Gestaltbarkeit. *Informatik Spektrum* 33(6), pp. 626–633

Owen, R.; Bessant, J.; Heintz, M. (eds) (2013): *Responsible Innovation: Managing the Responsible Emergence of Science and Innovation in Society*. London: Wiley

Pade, C.; Giese, B.; Koenigstein, S.; Wigger, H.; von Gleich, A. (2014): Characterizing synthetic biology through its novel and enhanced functionalities. In: Giese, B.; Pade, C.; Wigger, H.; von Gleich, A. (eds.): *Synthetic Biology: Character and Impact*. Heidelberg: Springer, pp. 71–104

Parens, E.; Johnston, J.; Moses, J. (2009): *Ethical Issues in Synthetic Biology: An Overview of the Debates*. Washington, DC: Woodrow Wilson International Center for Scholars

Paslack, R.; Ach, J.; Lüttenberg, B.; Weltring, K.-M. (eds.) (2012): *Proceed with Caution. Concept and Application of the Precautionary Principle in Nanobiotechnology*. Münster: LIT

Pedrono, E.; Durukan, A.; Strbian, D.; Marinkovic, I.; Shekhar, S.; Pitkonen, M. (2010): An optimized mouse model for transient ischemic attack. *Journal of Neuropathological Experimental Neurology* 69(2), pp. 188–195

Pellé, S.; Reber, B. (2015): Responsible innovation in the light of moral responsibility. *Journal on Chain and Network Science* 15(2), pp. 107–117

Perelmann, C. (1967): *Justice*. New York, NY: Random House

Persson, E. (2017): What are the core ideas behind the precautionary principle? *Science of the Total Environment* 557, pp. 134–141

Peterson, M. (2007): The precautionary principle should not be used as a basic for decision-making. *EMBO reports* 8, pp. 10–14

Plessner, H. (1928): *Die Stufen des Organischen und der Mensch. Einleitung in die philosophische Anthropologie*. Berlin: de Gruyter

Poser, H. (2016): *Homo Creator. Technik als philosophische Herausforderung*. Wiesbaden: Springer

Presidential Commission for the Study of Bioethical Issues (2010): New Directions: The Ethics of Synthetic Biology and Emerging Technologies.

https://bioethicsarchive.georgetown.edu/pcsbi/synthetic-biology-report.html [Accessed 21 July 2020]

Proudfoot, C.; Carlson, D. F.; Huddart, R.; *et al.* (2015): Genome edited sheep and cattle. *Transgenic Research* 24, pp. 147–153

Psarros, N. (2006): Diskussionsbeitrag. *Erwägen Wissen Ethik (EWE)* 17, pp. 594–596

Qiu, L. (2006): Mighty mouse. *Nature* 444, pp. 814–816

Qiu, R. Z. (2016): Debating ethical issues in genome editing technology. *Asian Bioethical Review* 8(4), pp. 307–326

Radder, H. (2010): Why technologies are inherently normative. In: Meijers, A. (ed.): Philosophy of Technology and Engineering Sciences. Vol. 9, Amsterdam: Elsevier, pp. 887–922

Rath, J. (2018): Safety and security risks of CRISPR/Cas9. In: Schroeder, D.; Cook, J.; Hirsch, F.; Fenet, S.; Muthuswamy, V. (eds.): *Ethics Dumping. Case Studies from North-South Research Collaborations.* Berlin: Springer, pp. 107–113

Rawls, J. (1993): *Political Liberalism.* New York: Columbia University Press

Reggia, J. (2013): The rise of machine consciousness: Studying consciousness with computational models. *Neural Networks* 44, pp. 112–131. https://doi.org/10.1016/j.neunet.2013.03.011

Reich, J. (ed.) (2015): *Genomchirurgie beim Menschen – zur verantwortlichen Bewertung einer neuen Technologie: Eine Analyse der Interdisziplinären Arbeitsgruppe Gentechnologiebericht.* Berlin: Berlin-Brandenburgische Akademie der Wissenschaften

Reuter, K. (2007): Tierzucht für den ökologischen Landbau – Probleme, offene Fragen, Lösungsansätze. In: Zukunftsstiftung Landwirtschaft (ed.): *Tierzucht für den Ökologischen Landbau – Anforderungen, Ergebnisse, Perspektiven.* Bochum: Zukunftsstiftung Landwirtschaft, pp. 8–9. http://orgprints.org/15131/1/reuter-etal-2007-Tagungsband_Tierzucht_Kassel.pdf [Accessed 15 Sep. 2020]

Ridley, M. (2003): *Nature via Nurture: Genes, Experience, & What Makes Us Human.* New York, NY: Harper Collins

Rip, A.; Misa, T.; Schot, J. (eds.) (1995): *Managing Technology in Society.* London: Pinter

Roberge, J. (2011): What is critical hermeneutics? *Thesis Eleven* 106(1), pp. 5–22

Robert, J.S.; Baylis, F. (2003): Crossing species boundaries. *The American Journal of Bioethics* 3(3), pp. 1–13

Robischon, M. (2007): Von Fischen und Genen. *Amazonas* 14, pp. 64–68

ROBOTS (2020): Pepper. https://robots.ieee.org/robots/pepper/ [Accessed 21 Sep. 2020]

Roco, M. C.; Bainbridge, W. S. (eds.) (2002): *Converging Technologies for Improving Human Performance.* Arlington, VA: National Science Foundation

Rodríguez-Rodríguez, D. R.; Ramírez-Solís, R.; Garza-Elizondo, M. A.; Garza-Rodríguez, M. D. L.; Barrera-Saldaña, H. A. (2019): Genome editing: A perspective on the application of CRISPR/Cas9 to study human diseases (Review). *International Journal of Molecular Medicine* 43, pp. 1559–1574. https://doi.org/10.3892/ijmm.2019.4112

Rohbeck, J. (1993): *Technologische Urteilskraft. Zu einer Ethik technischen Handelns.* Frankfurt: Suhrkamp

Rosemann, A.; Balen, A.; Nerlich, B.; *et al.* (2019): Heritable genome editing in a global context: National and international policy challenges. *Hastings Center Report* 49(3), pp. 30–42

Roskies, A. L. (2013): The neuroscience of volition. In: Clark, A.; Kiverstein, J.; Viekant, T.(eds.): *Decomposing the Will.* Oxford: Oxford University Press

Ruggiu, D. (2018): *Human Rights and Emerging Technologies: Analysis and Perspectives in Europe.* Foreword by Roger Brownsword. Singapore: Jenny Stanford Publishing

Ruggiu, D. (2019): Inescapable frameworks: Ethics of care, ethics of rights and the responsible research and innovation model. *Philosophy of Management* 19, pp. 237–265. https://doi.org/10.1007/s40926-019-00119-8

Ryder, R. D. (2000): *Animal Revolution: Changing Attitudes Towards Speciesism.* Oxford: Berg

Sand, M. (2018): *Futures, Visions, and Responsibility. An Ethics of Innovation.* Wiesbaden: Springer

Sandel, M. J. (2007): *The Case Against Perfection. Ethics in the Age of Genetic Engineering.* Cambridge, MA: Harvard University Press

Sarewitz, D.; Karas, T. H. (2006): *Policy Implications of Technologies for Cognitive Enhancement.* Phoenix, AZ: Arizona State University

Sato, K.; Sasaki, E. (2018): Genetic engineering in nonhuman primates for human disease modeling. *Journal of Human Genetics* 63(2), pp. 125–131

Sauter, A.; Gerlinger, K. (2011): *Pharmacological and technical interventions for improving performance: Perspectives of a more widespread use in medicine and daily life ("Enhancement")*. Berlin: TAB. https://www.tab-beim-bundestag.de/en/research/u141.html [Accessed 15 Sep. 2020]

Sauter, A.; Albrecht, S.; van Doren, D.; König, H.; Reiß, T.; Trojok, R. (2015): *Synthetische Biologie – die nächste Stufe der Biotechnologie*. Berlin: TAB. https://www.tab-beim-bundestag.de/de/untersuchungen/u9800.html [Accessed 15 Sep. 2020]

Savulescu, J. (2001): Procreative beneficence: Why we should select the best children. *Bioethics* 15(5/6), pp. 413–426. https://doi.org/10.1111/1467-8519.00251

Savulescu, J.; Bostrom, N. (eds.) (2009): *Human Enhancement*. Oxford: Oxford University Press

Schäfer, L. (1993): *Das Bacon-Projekt*. Frankfurt am Main: Suhrkamp

Schlachetzki, A. (1993): Künstliche Intelligenz und ihre technisch-physikalische Realisierung. In: Verein Deutscher Ingenieure (ed.): *Künstliche Intelligenz: Leitvorstellungen und Verantwortbarkeit*. VDI-Report, Vol. 17, Düsseldorf: VDI Verlag, pp. 72–82

Schmid, G.; Ernst, H.; Grunwald, A.; *et al.* (2006): *Nanotechnology – Perspectives and Assessment*. Berlin *et al.*: Springer

Schmidt, K. (2008): *Tierethische Probleme der Gentechnik: Zur moralischen Bewertung der Reduktion wesentlicher tierlicher Eigenschaften*. Paderborn: Mentis

Schwille, P. (2011). Bottom-up synthetic biology: Engineering in a tinkerer's world. *Science*. 333 (6047), pp. 1252–1254. https://doi.org/10.1126/science.1211701

Scott, N. R. (2005): Nanotechnology and animal health. *Revue scientifique et technique (International Office of Epizootics)* 24(1), pp. 425–432

Scott, S. (2006): Converting thoughts into action. *Nature* 442, pp. 141–142

Searle, J. R. (1980): Minds, brains, and programs. *Behavioral and Brain Sciences* 3(3), pp. 417–457

Selgelid, M. (2007): Ethics and drug resistance. *Bioethics* 21, pp. 218–221

Sethe, S. (2007): Nanotechnology and life extension. In: Allhoff, F.; Lin, P.; Moor, J.; Weckert, J. (eds.): *Nanoethics: The Ethical and Social Implications of Nanotechnology*. Hoboken, NJ: Wiley, pp. 353–365

Sharp, P. A. (2005): 1918 flu and responsible science. *Science* 310(5745), p. 17

Shriver A. (2009): Knocking out pain in livestock: Can technology succeed where morality has stalled? *Neuroethics* 2, pp. 115–124

Shriver, A.; McConnachie, F. (2018): Genetically modifying livestock for improved welfare: A path forward. *Journal of Agricultural and Environmental Ethics* 31(2), pp. 161–180

Shriver, A. (2015): Would the elimination of the capacity to suffer solve ethical dilemmas in experimental animal research? *Current Topics in Behavioral Neuroscience* 19, pp.117–132

Shumyatsky, G. P.; Malleret, G.; Shin, R. M.; *et al.* (2005): Stathmin, a gene enriched in the amygdala, controls both learned and innate fear. *Cell* 123, pp. 697–709

Siep, L. (2005): Enhancement, cloning, and human nature. In: Nimtz, C.; Beckermann, A. (eds.): *Philosophy – Science – Scientific Philosophy.* Paderborn: Mentis, pp. 191–203

Siep, L. (2006): Die biotechnische Neuerfindung des Menschen. In: Abel, G. (ed.): *Kreativität.* Akten des XX. Deutschen Kongresses für Philosophie. Kolloquienbeiträge. Hamburg: Felix Meiner Verlag, pp. 306–323

Simakova, E.; Coenen, C. (2013): Visions, hype, and expectations: A place for responsibility. In: Owen, R.; Bessant, J.; Heintz, M. (eds.): *Responsible Innovation.* London: Wiley, pp. 241–266

Singer, P. ([1975] 1990): *Animal Liberation.* New York, NY: Random House

Siune, K.; Markus, E.; Calloni, M.; Felt, U.; Gorski, A.; Grunwald, A.; Rip, A.; de Semir, V.; Wyatt, S. (2009): *Challenging Futures of Science in Society.* MASIS Expert Group. Brussels: European Commission

Sniderman, B.; Mahto, M.; Cotteleer, M. J. (2016): *Industry 4.0 and manufacturing ecosystems: Exploring the world of connected enterprises.* New York, NY: Deloitte

SoftBank Robotics (n.d.): Pepper. https://www.softbankrobotics.com/emea/en/pepper [Accessed 6 March 2020]

Sorgner, S. L. (2016): Three transhumanist types of (post)human perfection. In: Hurlbut, J. B.; Tirosh-Samuelson, H. (eds.): *Perfecting Human Futures: Transhuman Visions and Technological Imaginations.* Wiesbaden: Springer, pp. 141–158

Stachowiak, H. (1970): Grundriß einer Planungstheorie. *Kommunikation* VI(1), pp. 1–18

Stahl, B.; Eden, G.; Jirotka, M. (2013): Responsible research and innovation in information and communication technology: Identifying and engaging with the ethical implications of ICTs. In: Owen, R.; Bessant, J.; Heintz, M. (eds.): *Responsible Innovation: Managing the Responsible Emergence of Science and Innovation in Society.* Chichester: Wiley, pp. 199–218

Sternberg, E. J. (2007): *Are You a Machine? The Brain, the Mind, And What It Means to Be Human*. Amherst, NY: Prometheus Books

Steusloff, H. (2001): Roboter, soziale Wesen, ... In: Kornwachs, K. (ed.): *Conference Report*. Karlsruhe: Gesellschaft für Systemforschung, p. 7.

Stieglitz, T. (2006): Neuro-technical interfaces to the central nervous system. *Poiesis & Praxis* 4(2), pp. 95–109

Su, S.; Hu, B.; Shao, J.; *et al.* (2016): CRISPR-Cas9 mediated efficient PD-1 disruption on human primary T cells from cancer patients. *Scientific Reports* 6, 20070, pp. 1–13

Sunstein, C. (2005): *The Laws of Fear: Beyond the Precautionary Principle*. Cambridge: Cambridge University Press

Swierstra, T.; Rip, A. (2007): Nano-ethics as NEST-ethics: Patterns of moral argumentation about new and emerging science and technology. *NanoEthics* 1, pp. 3–20

Synbiology (2005): *Synbiology: An Analysis of Synthetic Biology Research in Europe and North America*. European Comission Framework Programme 6 reference contract 15357 (NEST). http://www2.spi.pt/synbiology/documents/SYNBIOLOGY_Literature_And_Statistical_Review.pdf [Accessed 3 May 2015]

Synth-Ethics (2011). Ethical and regulatory issues raised by synthetic biology. http://synthethics.eu/ [Accessed 3 May 2015]

Talwar, S. K.; Xu, S.; Hawley, E. S.; *et al.* (2002): Behavioural neuroscience: Rat navigation guided by remote control. *Nature* 417, pp. 37–38

Taylor, C. (1989): *Sources of the Self. The Making of the Modern Identity*. Cambridge: Cambridge University Press

TechRepublic. (2016): Pepper the robot: The smart person's guide. https://www.techrepublic.com/article/pepper-the-robot-the-smart-persons-guide/ [Accessed 21 Sep. 2020]

Thomson, J. (1985): The trolley problem. *The Yale Law Journal* 94(6), pp. 1395–1415. https://doi.org/10.2307/796133

Tisato, V.; Cozzi, E. (2012): Xenotransplantation: An overview of the field. *Methods in Molecular Biology* 885, pp. 1–16. https://doi.org/10.1007/978-1-61779-845-0_1

Truin, M.; van Kleef, M.; Verboeket, Y.; Deumens, R.; Honig, W.; Joosten, E. A. (2009): The effect of spinal cord stimulation in mice with chronic neuropathic pain after partial ligation of the sciatic nerve. *Pain* 145(3), pp. 312–318

Tucker, J. B.; Zilinskas, R. A. (2006): The promise and perils of synthetic biology. *The New Atlantis* 12, pp. 25–45

UNESCO (1997): Universal Declaration on the Human Genome and Human Rights. https://en.unesco.org/themes/ethics-science-and-technology/human-genome-and-human-rights [Accessed 19 May 2020]

University of Freiburg (n.d.): Engineering life: An interdisciplinary approach to the ethics of synthetic biology. http://www.egm.uni-freiburg.de/forschung/projektdetails/SynBio [Accessed 21 Sep. 2020]

van Baalen, S.; Gouman, J.; Verhoef, P. (2020): *Discussing the Modification of Heritable DNA in Embryos*. The Hague: Rathenau Institute

van den Belt, H. (2009): Playing God in Frankenstein's footsteps: Synthetic biology and the meaning of life. *Nanoethics* 3, 257. https://doi.org/10.1007/s11569-009-0079-6

van den Hoven, J.; Doorn, N.; Swierstra, T.; Koops, B.-J.; Romijn, H. (eds.) (2014): *Responsible Innovation 1. Innovative Solutions for Global Issues*. Dordrecht: Springer

van de Poel, I. (2009): Values in engineering design. In: Meijers, A. (ed.): *Philosophy of Technology and Engineering Sciences*. Vol. 9, Amsterdam: Elsevier, pp. 973–1006

van Est, R. (2014): *Intimate Technology. The Battle for Our Body and Behavior*. The Hague: Rathenau Institute

VDI – Verein Deutscher Ingenieure (1991): *Richtlinie 3780 Technikbewertung: Begriffe und Grundlagen*. Düsseldorf: VDI. English version: *Technology Assessment: Concepts and Foundations*. Düsseldorf: VDI

Velliste, M.; Perel, S.; Spalding, M. C.; Whitford, A. S.; Schwartz, A. B. (2008): Cortical control of a prosthetic arm for self-feeding. *Nature* 453, pp. 1098–1101

Vigen, T. (2015): *Spurious Correlations*. New York, NY: Hachette. https://www.tylervigen.com/spurious-correlations [Accessed 20 July 2020]

von Gleich, A.; Pade, C.; Petschow, U.; Pissarskoi, E. (2007): *Bionik: Aktuelle Trends und zukünftige Potenziale*. Bremen: Universität Bremen

von Gleich, A. (2020): Steps towards a precautionary risk governance of SPAGE technologies including gene drives. In: von Gleich, A.; Schröder, W. (eds.): *Gene Drives as Tipping Points*. Cham: Springer

von Schomberg, R. (2005): The precautionary principle and its normative challenges. In: Fisher, E.; Jones, J.; von Schomberg, R. (eds.): *The Precautionary Principle and Public Policy Decision Making*. Cheltenham: Edward Elgar, pp. 141–165

von Schomberg, R.; Hankins, J. (eds.) (2019): International Handbook on Responsible Innovation: A Global Resource. Cheltenham: Edward Elgar. https://doi.org/10.4337/9781784718862.00031

Waddington, I.; Smith, A. (2008): *An Introduction to Drugs in Sport*. London: Routledge

Wagner, P. (2005): Nanobiotechnology. In: Greco, R.; Prinz, F. B.; Lane, R. (eds.): *Nanoscale Technology in Biological Systems*. Boca Raton, FL: CRC Press, pp. 39–55

Wallach, W.; Allen, C. (2009): *Moral Machines: Teaching Robots Right from Wrong*. New York: Oxford University Press

Walton, D. (2017): The slippery slope argument in the ethical debate on genetic engineering of humans. *Science and Engineering Ethics* 23(6), pp. 1507–1528. https://doi.org/10.1007/s11948-016-9861-3

Walzer, M. (1983): *Spheres of Justice: A Defence of Pluralism and Equality*. Oxford: Basil Blackwell

Warneken, F.; Tomasello, M. (2006): Altruistic helping in human infants and young chimpanzees. *Science* 311(5765), pp. 1301–1303

Weber, K. (2015): Is there anybody out there? On our disposition and the (pretended) inevitableness to anthropomorphize machines. In: Gutmann, M.; Decker, M.; Knifka, J. (eds.): *Evolutionary Robotics, Organic Computing and Adaptive Ambience*. Vienna: LIT, pp. 107–122

Weckert, J.; Moor, J. (2007): The precautionary principle in nanotechnology. In: Allhoff, F.; Lin, P.; Moor, J.; Weckert, J. (eds.): *Nanoethics: The Ethical and Social Implications of Nanotechnology*. Hoboken, NJ: Wiley, pp. 133–146

Wei, J.; Wagner, S.; Maclean, P.; *et al.* (2018): Cattle with a precise, zygote-mediated deletion safely eliminate the major milk allergen beta-lactoglobulin. *Scientific Reports* 8, 7661: https://doi.org/10.1038/s41598-018-25654-8

Wendemuth, A.; Biundo, S. (2012): A companion technology for cognitive technical systems. In: Esposito, A., Esposito, A. M.; Vinciarelli, A.; Hoffmann, R.; Müller, V. C. (eds.): *Cognitive Behavioral Systems. Lecture Notes in Computer Science*. Vol. 7403, Berlin: Springer, pp. 89–103

Whitehouse, D. (2003): GM fish glows in the bowl, BBC news by June, 27, 2003. news.bbc.co.uk/2/hi/science/nature/3026104.stm [Accessed 16 Sep. 2020]

Williams, E.; Frankel, M. S. (2006): *Good, Better, Best: The Human Quest for Enhancement*. Summary Report of an Invitational Workshop. Convened by the Scientific Freedom, Responsibility and Law Program. American Association for the Advancement of Science

Wilson, E. O. (1999): *Consilience: The Unity of Knowledge*. New York, NY: Vintage

Winner, L. (1982): *Autonomous Technology. Technics-out-of-Control as a Theme in Political Thought.* Cambridge, MA: MIT Press

Winter, G. (2014): The regulation of synthetic biology by EU law: Current state and prospects. In: Giese, B.; Pade, C.; Wigger, H.; von Gleich, A. (eds.): *Synthetic Biology. Character and Impact.* Heidelberg: Springer, pp. 213–234

Wippermann, A.; Campos, M. (2016): *Genome Editing Technologies. The Patient Perspective.* London: Genetic Alliance UK

Witten, I. H.; Frank, E.; Hall, M. A. (2011): *Data Mining: Practical Machine Learning Tools and Techniques.* Burlington, MA: Morgan Kaufmann

Woese, C. R. (2004): A new biology for a new century. *Microbiology and Molecular Biology Reviews* 68(2), pp. 173–186

Wolbring, G. (2006): The Triangle of Enhancement Medicine, Disabled People, and the Concept of Health: A New Challenge for HTA, Health Research, and Health Policy. HTA Initiative #23. Alberta Heritage Foundation for Medical Research, Edmonton, Alberta, Canada. https://www.ihe.ca/advanced-search/the-triangle-of-enhancement-medicine-disabled-people-and-the-concept-of-health-a-new-challenge-for-hta-health-research-and-health-policy [Accessed 16 Sep. 2020]

Wolbring, G. (2008a): Oscar Pistorius and the future nature of Olympic, Paralympic and other sports. *Scripted* 5(1), pp. 139–160

Wolbring, G. (2008b): Why NBIC? Why human performance enhancement? *The European Journal of Social Science Research* 21, pp. 25–40

Wolbring, G. (2008c): The politics of ableism. *Development* 51, pp. 252–258. https://doi.org/10.1057/dev.2008.17

Wolbring, G. (2015): Gene editing: Govern ability expectations. *Nature* 527(7579), p. 446

Woll, S. (2019): On visions and promises: Ethical aspects of in vitro meat. *Emerging Topics in Life Sciences* 3(6), pp. 753–758

Yuyun, X.; Qiang, Y.; Jun, R. (2016): Application of CRISPR/Cas9 mediated genome editing in farm animals. *Hereditas (Beijing)* 38, pp. 217–226. https://doi.org/10.16288/j.yczz.15-398

Zhang, Y.; Massel, K.; Godwin, I. D.; Gao, C. (2018): Applications and potential of genome editing in crop improvement. *Genome Biology* 19, p. 210. https://doi.org/10.1186/s13059-018-1586-y

Zhaolin, S.; Ming, W.; Shiwen, H.; *et al.* (2018): Production of hypoallergenic milk from DNA-free beta-lactoglobulin (BLG) gene knockout cow using zinc-finger nucleases mRNA. *Scientific Reports* 8, 15430. https://doi.org/10.1038/s41598-018-32024-x

Index

ableism 295
accountability 32, 51, 94, 99, 129, 260, 268, 270, 273, 292, 299
action 36, 37, 40, 44, 49–54, 57, 82, 161, 242, 244, 255, 260
  communicative 38
  governmental 96
  human 21, 153, 190, 243, 260
  manual 74
  medical 121
  political 83
  protective 83
  unethical 296
action strategies 45, 242
actors 36, 37, 41, 44, 45, 50–52, 98–100, 113, 115, 117, 129, 130, 158, 159, 218, 223, 224, 260, 273, 275
  concrete 50
  groups of 45, 270
  moral 93
  policy 150
  real-world 39
  responsible 132
ADHD *see* attention deficit hyperactivity disorder
agenda-setting 47, 97, 98, 100, 171, 275, 299
aging 178–180, 196, 211
agriculture 2, 108, 109, 113, 115, 118, 120, 129–131, 143
algorithm 239, 242, 248–250, 253, 262, 264, 265, 271, 274, 275, 285, 295, 298
ambient assisted living 232
AML *see* animal microencephalic lump
ancestry 294–297

android Sophia 256
animal 1, 3–6, 8, 48, 88, 89, 107–137, 142, 144, 256–259, 279, 281–285, 287, 292, 296, 297, 302
  artificial 88
  breeding 8, 108, 138
  cloning of 141
  experimental 108, 111, 112, 118, 125
  farm 108–110, 120, 125, 126, 133, 144
  hypothetical 125
  non-enhanced 130
  sport 110
  transgenic 141
  utility of 109, 113, 119, 130
animal enhancement 4–6, 8, 11, 30, 89, 107–109, 113–115, 117–124, 127, 129–135, 144
animal experiments 43, 89, 111, 112, 122, 123, 132, 134, 151, 160, 165, 175
animal microencephalic lump (AML) 124–126
animal protection legislation 118
animal welfare 32, 47, 52, 115, 117, 118, 121, 125, 131, 133, 134, 294
Anthropocene 6, 277, 279, 282, 298
anthropomorphic language 10, 256, 301
approach 14–16, 20, 22, 35, 37, 44, 63, 64, 67, 68, 81, 82, 85, 143, 181, 182, 214, 215, 278,
  Aristotelian 54
  Baconian 278, 284, 287

bionic 103
consequentialist 56, 72, 79, 299, 300
deontological 54
hermeneutic 98, 102, 299
Jonas' 82
Kantian 53
manipulative 89, 288
methodological 37
precautionary 82, 164
Rawlsian 54
reductionist 68
reflexive 154
utilitarian 54, 55
Western 298
argument 79, 80, 83–85, 93, 102, 128, 129, 131, 150, 153–156, 198, 201, 203, 205–207, 264, 266, 296, 297
Habermas' 156, 207
argumentation 20, 40, 82, 196, 199, 200, 207, 209, 217, 266, 297
ethical 199, 207, 249
moral 39
normative 36
Aristotelian distinction 21, 29, 282
artificial companions 10, 230–234
artificial hippocampus 222
artificial intelligence 9, 229, 230, 232, 234, 236, 238, 240, 242, 244, 246, 248, 250, 252, 254, 276, 277
artificial limbs 174, 185
Asian traditions 165
attention deficit hyperactivity disorder (ADHD) 185, 186
automated vehicle (AV) 239, 240, 270
automation 48, 212, 235, 240
autonomous system 32, 235, 236, 240, 244, 271, 288

autonomous technology 2–4, 6, 9, 11, 229, 230, 233, 235–239, 241–243, 245–247, 250, 253, 255–263, 266–268, 270–276, 300
autonomous vehicle 230, 239, 240, 247, 250, 252, 254, 266–269
AV *see* automated vehicle

Baconian optimism 283
Baconian paradigm 298, 299
Baconian vision 102
Big Data 242, 253, 271
bioethics 3, 5–7, 86, 87, 122, 124, 149, 151, 152, 167, 184, 246
biological weapon 74, 77–79, 99
biosensor 66, 72, 73, 109
brain 111, 112, 170, 173–177, 188, 194, 221, 222, 224, 227, 260, 263, 264, 284, 286

cat 88, 112, 121, 257, 258, 293–296
artificial 88, 128, 257, 258, 293–295
natural 128, 293, 294, 296
chimeras 127, 128
Chinese room 264, 265, 294
Chinese twins 141, 163
cognitive enhancement 177, 196, 213, 216
conflicts 36–38, 41, 42, 44, 47, 53, 87, 115, 117, 119, 130, 131, 246
consciousness 113, 125, 173, 179, 230, 255, 266
consequentialism 7, 49, 54, 56, 94, 132, 209
consequentialist paradigm 56, 59, 68, 152, 300
converging technology 8, 9, 109, 113, 127, 171, 174, 196, 199, 219, 225, 229

cosmetic surgery 110, 121, 178, 181, 188–191, 198, 202, 209, 219
CRISPR-Cas9 138, 139, 143, 145–147, 155, 281
cyborg 204, 215, 217, 218
cystic fibrosis 145, 146

decision 36, 37, 40–42, 44, 50–54, 57, 116, 119, 161–164, 197, 200, 248–250, 253–255, 271–273, 275
  autonomous 249
  dilemmatic 248
decision-making 3, 5, 36–39, 42, 45, 82, 86, 154, 159, 161, 260, 264
determinism 166, 167, 261
digitalization 7, 9, 10, 174, 179, 220, 225, 227, 229, 230, 239, 242, 271, 274, 279
dignity 80, 86, 88, 91, 93, 126, 128, 157, 255, 257, 292
dilemma 32, 78, 79, 160, 247, 248, 292
disease 2, 111, 115, 116, 120, 125, 140, 145, 146, 152, 156, 159, 160, 162
  Alzheimer's 47, 111, 222
  genetic 143, 145, 162
  human 111, 144, 145
  Huntington's 112, 145
  infectious 147, 280
  monogenic 148
  Parkinson's 112
  sickle cell 145
DNA 62, 103, 142–146, 157, 160, 167, 291, 303

ELSA *see* ethical, legal, and social aspects
embryos 41, 48, 139, 146, 147, 149, 155–162, 164, 166, 200
  earliest stage of 147, 158

  edited 157, 163
  human 48, 138
  moral status of 33, 43
enhancement technology 2, 123, 193, 196, 197, 200, 202, 207–211, 213
ethical, legal, and social aspects (ELSA) 62, 69, 71
ethical analysis 33, 35, 46, 54, 199
ethical challenges 79, 81, 83, 85, 87, 89, 91, 93, 117, 119, 121, 123, 125, 246, 290
ethical criteria 42, 54
ethical debate 9, 49, 61, 62, 124, 141, 147, 150, 169, 192, 195, 206, 207
ethical issues 8–10, 61, 62, 88, 92, 93, 118, 119, 137, 138, 220, 226, 246, 247, 249, 251–253, 255, 257
ethical reflection 3, 5, 35–37, 39–46, 53, 54, 56, 158, 160, 179, 190, 191, 198, 199, 220, 221, 223
ethics 7, 35–39, 44–47, 49, 61, 62, 72, 93, 101, 105, 122, 123, 128–130, 133, 135–137, 139, 140, 149, 195–198, 225, 300, 302
  animal 6, 89, 123, 124, 126, 129, 130, 293
  consequentialist 7, 54, 55, 62, 253
  deontological 249
  discourse 39, 209
  libertarian 216
  machine 32, 246
  normative 35
  pathocentric 32, 118, 124, 126, 259
  problem-oriented 36, 37, 39, 41, 43, 93
  professional 49

European enlightenment 24, 104,
181, 192, 236, 271, 278
evolution 3, 7, 18, 66, 67, 76, 77,
104, 137, 205, 206, 277, 278,
283, 287
artificial 67
natural 76, 104, 277, 278, 283,
287

fears 54, 59, 60, 81, 83, 89–93, 97,
101, 201, 205, 272, 275
apocalyptic 57, 215
heuristics of 82, 85
speculative 59
fertilization 139, 141, 146, 161
Fordist mass production 238
Frankenstein's monster 90

gene 63, 77, 83, 110, 139, 140,
143, 145, 148, 151, 161, 162,
166
gene editing 6, 8, 108, 109, 113,
118, 123, 135, 140, 143, 159,
160, 163, 165
gene therapy 77, 142–148
genetically modified organism
(GMO) 2, 48, 55, 61, 76, 86
genetic intervention 138, 147,
155, 201, 207
genome 9, 79, 110, 138, 139,
142–144, 147, 155, 157, 158,
162, 166, 167, 296
cell's 143
cow's 110
human 138, 152, 157, 166, 279,
284
genome editing 1, 2, 8, 9, 137–140,
142–145, 147–153, 155, 156,
158–160, 162, 163, 165–167,
284, 287
germline intervention 1, 9, 28, 33,
138, 141, 144, 147–152, 156,
158–161, 163–165

germline therapy 142, 146, 148,
149
GMO *see* genetically modified
organism

human autonomy 197, 200, 226,
230, 235, 236, 243, 250, 252,
253, 260, 263, 264, 270–276,
278
human driver 239, 240, 249,
251–254, 268, 270, 272
human enhancement 1, 4, 6, 9, 11,
58, 59, 107, 169–214,
216–222, 224–226, 228, 298,
300
humanity 33, 47, 103, 226, 227,
241, 278, 281, 300
human–machine interface 3, 10,
170, 229, 233, 247, 251, 268

informed consent 119, 151, 156,
162, 197, 200, 207, 209, 213,
217, 227
interventions 6, 8, 9, 46, 79–81,
108, 109, 120–123, 132, 135,
137–139, 146, 155–159,
161–163, 186, 187, 207, 216,
217
heritable 150
medical 147, 202
pharmacological 214
technological 204
veterinary 121

learning 18, 20, 30, 166, 170, 176,
211, 217, 218, 238, 243–245,
248
life 1–8, 10, 11, 13–20, 24–29,
31–33, 62–65, 67, 68, 70, 79,
80, 86–93, 101, 102, 104,
105, 135–137, 161–163, 196,
197, 200, 201, 246, 256, 257,
277–284, 286–303

artificial 67, 288, 290, 291, 294, 296, 301
autonomous 156, 232
dignity of 3, 52, 60, 101
fabrication of 29, 88, 289
ladder of 278, 293, 298, 301
miracle of 302, 303
natural 31, 67, 283, 284, 287
lifespan 112, 120, 162, 174, 177–179, 192, 199, 216
lifestyle 43, 162, 208, 210, 220
living entity 16–18, 24, 26, 29–31, 35, 277, 282, 285, 286, 291–296, 298, 300–303
living object 17–20, 31, 76, 78
living system 5, 7, 61, 63, 64, 69, 76, 79, 81, 86, 105, 229, 245, 246, 290

machine 18–21, 26, 27, 30–32, 62, 64, 81, 134–136, 172–175, 179, 227, 228, 235, 237, 238, 257, 258, 285, 286, 301
animated 134
autonomous 246, 252, 257, 261, 266, 267
inanimate 134
molecular 66
programmable 230
random 262, 263
washing 128, 278, 294
meat 125, 126, 130–133, 283
medicine 4, 44, 49, 72, 79, 102, 149, 151, 164, 178, 202
human 120, 138
regenerative 146
reproductive 141
veterinary 108, 120
metabolism 15, 18, 172, 288, 289, 302
model 10, 27, 30, 64, 103, 104, 111, 112, 144, 145, 192, 193, 242, 244, 245, 263, 264, 278, 279, 284–287, 296, 297

animal 111, 122, 123, 125, 145
EEE 51, 55, 62, 94, 96, 97, 153, 158, 159, 218, 223, 259, 300

nanorobots 75, 178, 289, 292
NEST *see* new and emerging sciences and technologies
neurocognitive enhancement 200, 201
neuro-electric interface 217, 222–224
neuro-implants 73, 175, 224
new and emerging sciences and technologies (NEST) 46, 57, 59, 96, 154
new Renaissance 171
normative framework 41–43, 51, 80, 118–120, 123, 124, 130, 186, 188–192, 217, 219, 223–225
normative uncertainty 9, 10, 33, 35, 37, 39–42, 46–50, 80, 81, 89, 93, 147, 158, 159, 188–191, 195, 196, 229, 230, 246

optimistic futurism 215
organism 8, 15, 19, 20, 46, 48, 62, 63, 69, 74, 75, 86–89, 142, 143, 145, 151, 158
artificial 80
laboratory-created 71
living 1, 2, 4, 6, 10, 15, 18, 26, 30, 61, 64, 87, 258, 263
natural 1
redesigned 72
synthetic 73, 81
orientation 3–5, 7, 29, 31, 33, 35–37, 49, 50, 54, 56, 58, 188–191, 198, 199, 202, 203, 247, 248, 278

pain 32, 87, 118, 119, 122, 124, 126, 127, 132, 272, 293, 296

patient 44, 77, 121, 158, 164, 185,
  186, 193, 194, 217, 219, 241,
  258
perspective 4, 5, 13, 14, 17, 18,
  22–25, 59, 60, 95–97, 104,
  105, 119, 135, 158, 181, 193,
  194, 208, 209, 214, 272, 273,
  275–277, 285, 286, 290, 291,
  303
  advocatory 115, 119
  biased 272
  consequentialist 57
  epistemological 291
  ethical 37, 199, 303
  hermeneutic 98, 301
  libertarian 209
  phenomenological 265, 294
  theological 70
  utilitarian 119
philosophical debate 183, 211,
  301
philosophy 1, 3–7, 10, 13–16, 35,
  38, 136, 137, 169, 171, 246,
  261, 262, 299, 302, 303
  academic 45
  continental 156
  explorative 199
principle 38, 55, 56, 60, 66, 67,
  82–86, 103, 104, 172, 214,
  217, 247, 249, 290, 294
  liberal 254
  precautionary 49, 55, 76, 77,
  81–86, 92, 93
  utilitarian 39
prosthesis 2, 73, 224, 227
public debate 45, 47, 49, 54, 70,
  91, 118, 138, 141, 150, 205

realms of life 225, 279
regulations 31, 33, 35, 38, 44, 100,
  110, 140, 142, 154–158, 163,
  165, 167, 213, 217, 218
  environmental 55
  gene 67

  legal 252
  technical 287
relationship 8, 39, 105, 108, 133,
  134, 136, 214, 217, 220, 223,
  224, 231, 233, 237
  doctor–patient 223
  functional 111
  fundamental 118
  technical 64
  technocratic 39
religion 15, 92, 102, 118, 128, 134,
  156, 278
reproducibility 26–28, 289
research 43, 44, 48, 49, 66, 70, 78,
  91, 93, 94, 96, 97, 99, 100, 109,
  121–123, 130, 132–134,
  140–142, 148–150, 153, 154,
  163, 164, 220, 221, 223, 224,
  235
  biomedical 122
  brain organoid 112
  clinical 149, 160, 164
  disease 145
  empirical 216, 220
  preclinical 159, 160
  scientific 48, 61, 78
  stem cell 41
research funding 44, 49, 99, 150,
  219
responsibility 2, 3, 7–11, 45, 46,
  49–53, 55–61, 68, 94, 96–101,
  128–132, 137, 138, 158, 159,
  161, 163–165, 216–219,
  259–261, 264, 266–270,
  273–275, 292, 293, 299–303
  disappearing 131, 273
  distribution of 8, 9, 32, 50–52,
  94, 164, 240, 267, 268, 270
  empirical dimension of 51
  ethical dimension of 52, 131
  ethics of 36, 49, 51, 53, 55, 57,
  59, 98, 292
  human 46, 226, 259, 273, 274,
  278, 299, 301

scientific 52, 219
responsibility assignment 50, 51,
  93, 94, 98, 100, 101, 129, 130,
  158, 160, 218, 221, 224, 267,
  270, 299, 300
responsibility considerations 129,
  131, 132, 159, 164, 259, 261,
  263, 265, 267, 269
responsibility issues 6, 9, 46, 50,
  53, 117, 130, 138, 148, 169,
  250
responsibility regimes 267–270
road traffic 240, 243, 247, 249,
  252, 254, 266, 269, 272
robot 2, 3, 5, 6, 31–33, 204,
  229–231, 233–235, 238,
  241–243, 245, 246, 251,
  255–266, 285, 289, 292–294,
  296–298, 300–303
  autonomous 3, 10, 30, 229, 243,
    257, 288, 289, 292, 299
  care 10, 43, 59, 230, 233, 240,
    242, 273, 275
  humanoid 31, 244, 289,
    295–297
  messenger 262, 265, 266
  service 274
  sex 233
  soccer-playing 243

salvation 97, 279
scenarios 54, 56, 57, 59, 73, 75, 76,
  78, 82, 101, 126, 206, 209
  adverse 271
  conceivable 247
  dilemma 248
  gray goo 76
  green goo 75
  horror 81
  market-oriented 209
self-driving car 226, 239, 247, 254,
  255, 270, 272, 273, 275
self-organization 2, 72–74, 76, 79,
  81, 82, 84, 101, 238, 294

smart factory 231, 237, 256, 257,
  266, 270
social life 27, 28, 233, 234, 253
social practice 36, 38–40, 42, 53,
  218, 267, 270
society 26–28, 33, 38–40, 46, 47,
  96, 97, 100, 101, 152–154,
  173, 174, 197, 198, 201, 208,
  210, 211, 213, 214, 216–218,
  220, 225, 227, 254, 255, 270,
  273, 274
  aging 240
  democratic 91
  digitalized 216, 253
  enhancement 198, 209–211,
    213, 215, 216
  modern 25, 37, 130, 255, 269
  performance-enhancing 210
  performance-oriented 210, 214
  secular 128
  totalitarian 227
species 74, 75, 88, 113, 119,
  121–123, 127, 128, 134, 204,
  206, 257, 258, 263, 283, 289
speculation 56, 82, 87, 148, 150,
  152, 163, 176, 179
sports doping 190, 191, 198
stakeholder 49, 51, 71, 95, 270
status 52, 53, 83, 114, 116, 125,
  128, 184, 185, 208, 214, 256,
  257, 262, 264, 272
  epistemological 52
  judicial 127
  methodological 16
  ontological 126
  social 249
suffering 118, 119, 122, 124–126,
  132, 133, 162, 283, 293, 296
synthetic biology 6–8, 11, 20, 29,
  30, 43, 57, 58, 61–65, 67–83,
  85–105, 279, 283, 284, 288,
  289, 291, 298–300, 302, 303

technology assessment 11, 21, 56,
  62, 69, 70, 141, 142, 150, 215,
  219, 269, 275
technology determinism 153, 250,
  274, 275
terrorists 77, 78, 243
therapy 46, 47, 111, 123, 145, 148,
  160, 184, 186, 263
  antimicrobial 146
  classical 178
  clinical 145
  complex 151
  somatic 147
trolley problem 240, 247

uncertainty 29, 31, 32, 47, 49, 52,
  55–58, 83–85, 89, 203, 235,
  236, 248, 300, 303

unintended side effects 4, 25, 143,
  154, 202, 207, 280, 287, 299

value judgments 59, 185, 281
values 38, 39, 43, 45, 86–88, 115,
  116, 118, 127, 129, 152, 153,
  240, 249, 274, 275, 278, 281,
  292, 293
virus 6, 75, 77–79, 87, 101, 280,
  282, 284, 288, 292, 294
visions 47, 58, 97, 99, 102, 103,
  133, 134, 136, 146–148, 150,
  171, 178, 179, 204, 205,
  213–215, 225, 232–234

Western culture 29, 282, 288

xenotransplantation 111, 123,
  127, 146

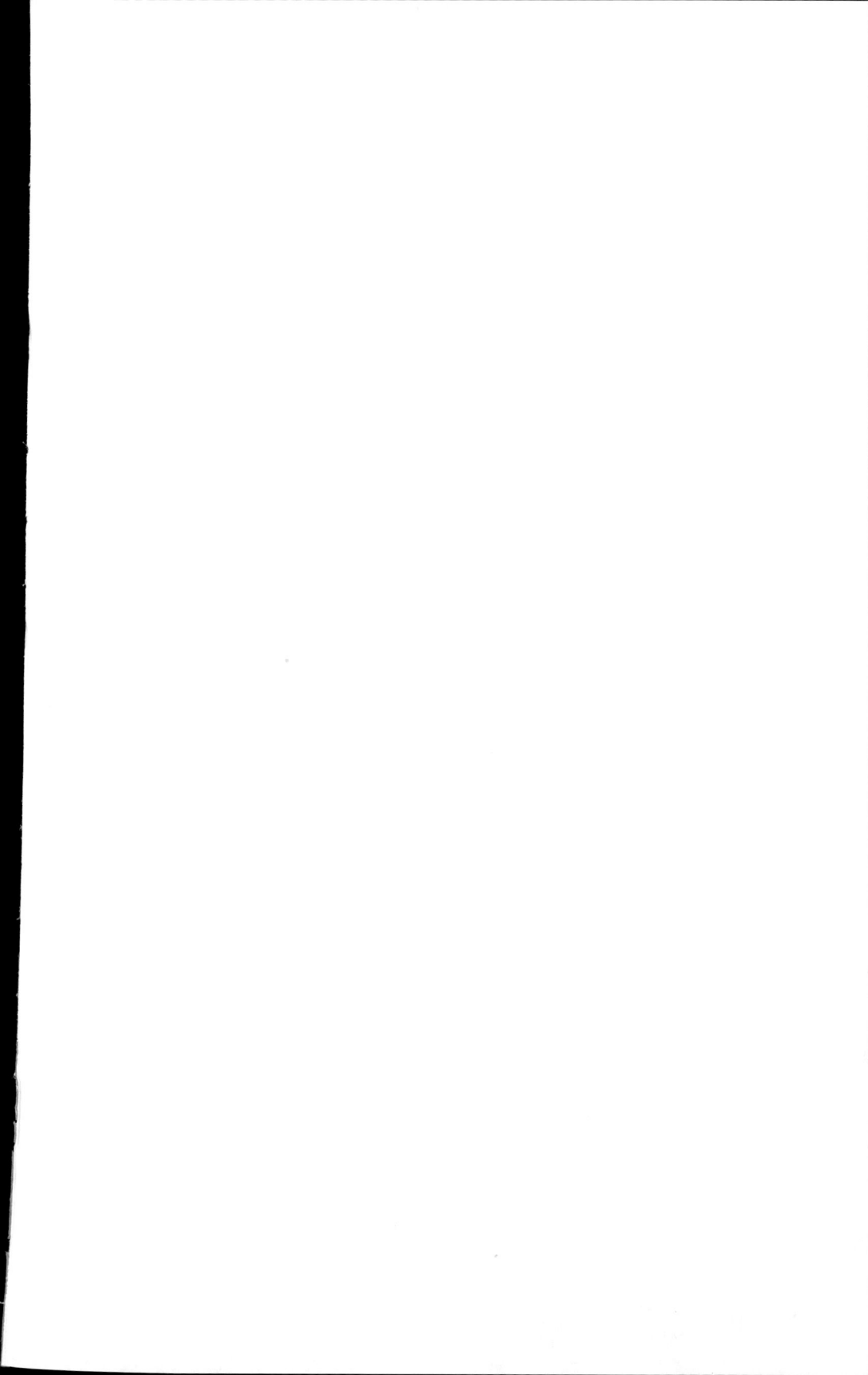

For Product Safety Concerns and Information please contact our EU
representative GPSR@taylorandfrancis.com
Taylor & Francis Verlag GmbH, Kaufingerstraße 24, 80331 München, Germany

www.ingramcontent.com/pod-product-compliance
Lightning Source LLC
Chambersburg PA
CBHW050454190326
41458CB00005B/1274